Farming in Yorkshire

60p.

By the same author:
THE CLEVELAND WAY
LYKE WAKE WALK

Of related interest:
THE FACE OF NORTH-WEST YORKSHIRE
PENNINE WALLS

Farming in Yorkshire

A Regional Survey

by Bill Cowley

DALESMAN BOOKS
1972

The Dalesman Publishing Company Ltd.,
Clapham (Via Lancaster), Yorkshire

First Published 1972

ISBN: 0 85206 130 7

Printed by Galava Printing Co. Ltd.
Hallam Road, Nelson, Lancs.

Contents

The cover illustrations show: — Front: Combine harvesting at Potto Hill. Back: Tan Hill sheep show (Geoffrey N. Wright). Other photographs are between pages 33 and 40, and 57 and 64.

The line drawings in the text are by Alec Wright; for the most part they depict Goulton Grange farm.

Preface

YORKSHIRE is an agricultural microcosm. In its 3,155,962 rural acres it has a greater variety of soil types and agricultural systems than have many countries. This book is one man's view of his own county, coloured by his own farming experience, and friendships, over half a century. A dozen people could write a dozen different books on the same subject. In one such volume it is impossible to be exhaustive; there are bound to be many omissions. I have tried to present a real picture of Yorkshire farming and the human stories behind it rather than a technical selection of the best Yorkshire farms. Yorkshire has always had some of the best farmers in the world, pioneers in land reclamation, new crops, new machinery, advanced breeding techniques. It has also been rich in characters, like Bobby Dowson of Bilsdale who was content to scrape a bare sustenance from his few dales acres, and lived only for hunting, or Douglas Charlesworth, a small dairy farmer who became a noted local historian and a fine dialect poet.

If this book gives some idea of the rich diversity of Yorkshire, both human and physical, I will be happy. A southern agricultural expert after some years in Yorkshire made this comment on the differences he found: "What strikes me about Yorkshire farming," he said, "is its soundness, its solidity, its continuity. Yorkshire farmers look at the whole farm—no flash in the pan schemes for them. They were not panicked out of livestock in the 1950s like some southern farmers were. They retained their grass or root breaks, their cattle and their sheep. The hill men have had to be hardy, but they have survived. Yorkshire farmers may be behind in processing and packing, in vertical integration and dealing with the big combines, but they are basically healthy, and whoever else may go to the wall you may depend on it that they will survive." Like their cricket team, Yorkshire farmers are dour professionals who do not suffer fools gladly. Their farming is a business which has to make brass, but it is much more than that. It is their life as well as their living, and the land, and the family farm, are seen and judged across several generations.

1. The Framework of Yorkshire Farming

THE Pennine hills and valleys, the North Yorkshire moors and dales, the river systems and the underlying rocks of the central plains, as affected by the movement of lakes and glaciers long before farming began, dictate the diverse pattern of Yorkshire farming today. Among them the three Ridings have everything from the boulder clay of Cleveland to the blow-away sands of Skipwith; the limestone uplands of Craven to the liassic shale valleys of Farndale and Rosedale; the magnesian limestone of Tadcaster to the rich warp of Ouse and Humber; the acid rhubarb soils of Leeds and the one-time liquorice soils of Pontefract to the barley-growing chalk wolds of the East Riding; and the stark fells of the northern Pennines to the old glacial lake of Pickering. Each smaller area in turn has its own diversity. Between Ouse and Derwent a field of clay can lie next to a field of sand, according to the spread of the meltwater. The clay may be boulder clay deposited by a glacier or the even stiffer lacustrine clay deposited by a lake. Such differences can dictate a farming policy between field and field.

On the Cleveland geological map the line of outcrop of the "Rhaetic beds" is shown as going through a seventeen acre field of mine at Potto Hill, near Stokesley. I am not sufficient of a geologist to know what the "Rhaetic beds" are, but I do know that that particular field is a difficult one to work, and is better not left too long in corn. Then where Potto Hill rises out of the lower boulder clay is one field where the moles bring up almost pure sand. On this field my predecessor grew the best turnips and swedes in the district —and made the best turnip wine I have ever tasted! On the varied soils along the lower reaches of Ouse, Wharfe and Derwent, where fingers of estuarine warp and alluvium reach out among the clays and the sands, there is great variety of agricultural landscape and settlement. Further west, where boulder clay lies over the dip slope of the magnesian limestone, pastoral parkland is mixed with malting barley and potatoes. On the heavier clay soils, where drainage is good, wheat is a very profitable crop; where drainage is bad there is permanent pasture, with cattle.

Villages are often sited on glacial moraines, and here and there in the heartlands of Yorkshire are remnants of ancient heaths and birchwoods on outwash sands, pockets of peat, or primeval bog—bits of the forests of Elmet, Galtres and Knaresborough. Probably our oldest land surface however is that of the North York Moors, which were never over-run by the glaciers that swept east from the Pennines or south from the Cheviots. When the valleys and plains were forest and swamp, they and the chalk wolds were the homes of the early races in Yorkshire and of the first primitive farmers, some four thousand years ago. There are probably fields on the wolds, or parts of fields, that have been in cultivation ever since. On the moors much of the older cultivation has gone back to peat and heather, though its traces can still be found. But there are odd farms even here, such as Sleddale on the Guisborough moors, where Celtic fields underlie the modern ones, and cultivation has been continuous for at least two thousand years.

Even the wild moors which have never been cultivated, like the fells of the Pennines, have an agricultural as well as sporting value, being grazed extensively by sheep as well as by grouse. All moors may seem the same to a passing motorist, but like the cultivated fields of the plains their value as grazing varies a great deal with the underlying rock structure. This may be well-drained sandstone, giving a dry moor with good stock carrying capacity. Where the rock is broken and uneven, the result may be a wet moor, with pockets of peat bog, much less valuable for grazing.

Cleveland is an excellent starting point for a closer look at the geological framework of Yorkshire farming, for here in a small area the effect of the different strata and soils can be seen in a startling way. From the Tees, along which are some alluvial soils, often lying on free-draining gravel, the Cleveland Plain of boulder clay dropped by the glaciers stretches south to the steep scarp of the Cleveland Hills. Here the land lifts suddenly to between 1,000 and 1,500 feet above sea level—from modern arable farming to a primitive grazing economy. Even on the scarp however you can see that the strata of sandstones are lying on top of softer shales—the Lias. From the scarp these strata slope gently down to the south, and the streams which flow south have worn through the sandstones into the lias, forming fertile valleys. Ten miles south again, these valleys suddenly narrow where the sandstones have dipped down to stream level, the scenery is wild again, and you would scarcely suspect the fertility higher up. At this point the ridges mount up suddenly in another scarp, where oolitic limestone over-lays the sandstones, forming the Tabular Hills—Hawnby and Byland, Roppa and Beadlam, Cropton and Cawthorne. Where the heather ridges have dropped to under 800 feet, the limestone plateaux rise to 1,000 or 1,100 feet again. But where they in turn slope south from the scarp, the soils on top of the limestone are fertile, and from large arable fields above you can look down onto the heather below.

Often the streams disappear in the limestone, emerging once more in the Vale of Pickering beyond. A vast glacial lake in the Ice Age, this is a rich and fertile tract beyond which again rise the Chalk Wolds. These are the northern end of the chalk ridge which comes up from the south coast diagonally across England, from the Chilterns to the Lincolnshire Wolds. These again slope down to the alluvial soils of the Humber in the south or to the complicated clays, glacial sands, gravels and old lake bottoms of Holderness in the east. Historically there has been some inter-relation between the farming in these various eastern districts. The moors and dales have been a reservoir of men and of stock. Cattle and sheep were reared on the poorer lands, and sold down to the Cleveland Plain or to the Vale of Pickering for fattening. Young men and women from the small farms of the dales found work on the large farms of the plain or the Wolds. A similar inter-dependence is seen on a larger scale in the rest of Yorkshire.

The Pennines, the backbone of England, form the western boundary of Yorkshire. In places, between the Howgill Fells and the Forest of Bowland, this boundary is only ten miles from the Irish Sea. Some Yorkshire valleys drain to the west and climate, soils and agriculture bear more relation to Lancashire and Westmorland than to the rest of Yorkshire. But from the main Pennine watershed, whose peaks rise to 2,591 feet at Mickle Fell, the great dales running east into the Vale of York form the basic framework of Yorkshire farming.

The geology of the Pennine hills and dales is by no means simple. Some of the oldest rocks, the slates, can be seen in places, in the south west of Craven and in Teesdale, and the rounded grassy Howgill Fells round Sedbergh are of slate, providing a contrast to the limestones and millstone grits of the main Pennines. The limestone steps, scars and terraces are a feature of much of the northern Pennines, particularly Craven, Malham and Wensleydale, giving good pasture for both sheep and cattle. It might be said that Wensleydale cheese is based on the "Yoredale" system of limestones. Where the Millstone Grit predominates, as in the Cross Fell–Mickle Fell block, and at the head of Swaledale, an acid moor of peat bogs, rough grass and heather is the result. This occurs again south of the limestones, in the industrial West Riding and towards the Derbyshire border. Grazing on this sort of fell is not so good. The shepherd likes a mixture of "white" (grassy) and "black" (heathery) moor, since the heather is a more reliable winter food for the hardy sheep which alone can survive here. The predominant breed on these moors is the Swaledale, with some Rough Fell and, towards the south, Lonks and Derbyshire Gritstones. In the limestone country the Dalesbred, slightly larger than the Swaledale, predominates. Thus the rock structure and soil type affect not only the type of farming followed, but the type of animal which has been bred.

Three other features of Dales farming are due to geology. The dry-stone walls which divide farm and field and fell have been built over the centuries, hundreds of miles of them. The stones which were cleared from the pastures provided shelter from the wind and boundaries for grazing. Dales farms tend to lie in strips, from the stream below to the ridge above, giving each farm some good bottom land for meadow, some medium grazing land, some rougher enclosed grazing won in previous centuries as an intake from the moor, and the open moor or fell where each has grazing rights, either as stints from the landlord or as commonly owned grazing. The third feature resulting from this lay-out is the number of isolated stone barns which are dotted about the various fields. These are usually 28 feet x 14 feet, built to hold from four to six cattle for the winter, and the hay to feed them on. In this way transport of hay and of manure was reduced to a minimum, and only the milk, if any, had to be carried to the main farm-house.

Neither the Cleveland Dales nor the Pennine Dales can be considered in agricultural isolation. Both were affected by industrial or semi-industrial intrusions—flax spinning and weaving in Cleveland, hand-knitting in the northern Pennine Dales, wool-weaving further south. The growth of the woollen industry in the West Riding led to a great demand for sheep for their wool rather than their meat. In the 18th and 19th centuries the staple diet of the mill workers of Bradford and Halifax was top quality oats bought from the Vale of York and Cleveland. Where Cleveland had its alum and later its ironstone mining industries to provide employment for surplus agricultural labour, the northern dales had their leadmining. Lead was mined in Nidderdale in Roman times, and many lead mining settlements were built in Swaledale, Arkengarthdale and Teesdale in the Middle Ages. In the 18th and 19th centuries lead mining was at its peak and many of the leadminers were also farmers. Through the toil of generations they reclaimed land from the moor and walled round their intakes and smallholdings at the moor edge. None of these were big enough to provide a living when lead mining ceased early in the present century. A few are still kept on by other part-time workers—postmen, roadmen and the like—but most lie derelict now and the moor once more is taking possession.

In their lower reaches the Pennine Dales broaden into fertile valleys before finally merging into the great Vale of York. It is generally below the 500 feet contour—below Grinton, Aysgarth, Pateley or Bolton Bridges—before corn becomes of any importance, though an odd field may be grown higher up. Below Otley, Ripley, Tanfield and Easby dales farming becomes that of the Vale of York. The Cleveland Plain and the Vales of Richmond, Catterick and Mowbray all join to form the northern part of the Central Plain or Vale of York. Here it is no more than fifteen to twenty miles wide, and from the deep heather of Scarth Wood Moor or Black Hambleton you can look right across this rich country to the Pennine foot-hills, and see the clefts in them that form the higher dales. Below Thirsk, however, the Vale widens rapidly, until south of York no hills of any size are in sight.

Over most of the Vale the underlying rocks are also lost to view, covered by many feet of glacial or riverain deposit. Here it is the glaciers and the rivers, and the quality of the drainage, that have the main effects on agriculture. There are some exceptions to this, the most notable being that separating the Millstone Grit and the Coal Measures of the industrial West Riding from the prevailing drift of the Vale is a belt of Magnesian Limestone. Between Tadcaster and Doncaster this provides some of the best-working arable land in the county, free draining, and famous for high quality potatoes, malting barley and peas. North of Boroughbridge this belt is mainly mixed with drift, but just east of Ripon is another stretch of excellent land, rich red sands sometimes referred to as "the golden mile" which can be worked at almost any time and will grow anything.

In many parts of the main Vale, clays and sands can be close together, with gravels and alluvial soils near the rivers. Several glacial moraines cross the plain, or leave stray hummocks of sand and gravel 50 feet above the general surface, and many of the villages are built on these—North Duffield, Skipwith and Thorganby on the Escrick moraine; Easingwold, Boroughbridge, Thirsk, and Thormanby. The long narrow ridge running north from Borrowby to Winton and East Harlsey is mainly of glacial origin and provided a good route for cattle and horses before the shorter turnpike road (later the A19) was built in 1804. The *Cat and Bagpipes Inn* at East Harlsey in fact refers to the "caterans" or Scottish drovers who used this road, though an even better route for cattle was the Hambleton Drove Road from Swainby to Sutton Bank and Crayke.

One strange result of another line of low morainic hillocks only a mile south of the Tees, from Great Smeaton to Hornby, is to turn away the drainage which would normally have flown into the Tees. Thus the river Wiske, which rises just above Swainby, flowing west to Staddlebridge, then north west to Appleton Wiske, is turned back south by these hills that rise only 50 feet above it. Though its bed is at that point only 150 feet above sea level, it has to flow all the way down through the Vale again, joining the Swale below Kirby Wiske, some fifteen miles to the south, and eventually reaching the sea by the Humber, after another 50 miles, a fall of less than a yard per mile, or about 1 in 2,000. Not unnaturally its route is marked on the geological map by a narrow belt of alluvium, broadening in those sections most liable to floods.

South of the Escrick and Askham Bryan moraines the deposits which form the Vale of York soil are lacustrine rather than glacial—they were deposited in immediate post-glacial times by a great "Lake Humber." Much of this remained marsh until the art of draining was developed in the 17th and 18th centuries. A Yorkshire poem of 1754 bewails the loss of Snaith Marsh on behalf of "Young Robin of the plain":—

But far more waeful still that luckless day
Which with the commons gave Snaith Marsh away;
Snaith Marsh our whole town's pride, the poor man's bread,
Where tho' no rent he paid, his cattle fed,
Fed on the sweetest grass which here rife grew,
Common to all, nor fence nor landmark knew.

Draining and enclosure led to tremendous improvements in agriculture, but the poorer villagers with no share in the enclosed lands suffered badly. Since draining, four main soil types can be differentiated, east of the magnesian limestone:

1. Lake bottom clays and heavy alluvial soil.
2. Lake shore sands and light alluvial soil.
3. Warps (formed by later river flooding and the deposit of silt, sometimes natural, sometimes artificial).
4. Peats—mostly near Thorne and Hatfield.

Underlying these deposits are two formations which occasionally crop out—Bunter sandstones to the west (this is the red sandstone found near Ripon, and near Snaith, and also forms Brayton Barff, the remarkable hillock south-west of Selby) and Keuper Marl east of this. Where the Lake Bottom or Lacustrine Clays and Heavy Alluvial Soils occur in conjunction with the Keuper Marl, the water table is high, the soil rich, and heavy yields of cash root crops are normally obtained, little affected by drought. On the contrary, sand lands, particularly if lying over sandstones, are very susceptible to drought. In the summer of 1970 such lands towards the Nottinghamshire border grew only 5 to 8 cwts. of barley to the acre instead of a normal 30 to 35 cwts. or more.

The warp soils are formed by layers of river silt. Many of them have been artificially created by letting the river water flood onto the fields at high tide and run out slowly through a system of sluices, leaving a layer of silt behind. The process was repeated over many months until a depth of two to three feet had been deposited. A still more laborious process was "cart-warping" which involved carrying warp away by horse and cart to dress a field away from the river. To put nine inches of warp on one acre would involve carrying 1,000 tons. Even tidal warping has long been uneconomic, but when it did take place, whether over clays, sands or peats, it made them into very productive land. A similar process used in the past to improve blow-away sands in the Vale of York was marling, by heavy clay or Keuper Marl carted from a pit. Usually marl is found under or close to the sand. "Malpit" is a frequent place-name. A dressing of 150 tons per acre used to be given profitably, but during the 1939–45 war this was estimated to cost almost £30 an acre and except perhaps on a small market-garden scale this also has long been uneconomic.

The best warp lands are down the Humber, as at Sunk Island, and much of this warping is natural. But higher up, particularly round the Derwent junction, much riverside land is too low to drain and is more of a liability than an asset. In the 17th century Vermuyden drained some of this type of land when he made a new channel for the river Don near Goole—the "Dutch River." This rather desolate area between Goole and Bawtry seems scarcely part of Yorkshire. In some places there are a thousand acres with no inhabited farm house, and the land is farmed by big agricultural firms from Goole. In one field between Hatfield Woodhouse and Blaxton the three counties of Yorkshire, Nottinghamshire and Lincolnshire meet together. Within a few miles of this can be seen strong clay, unworkable unless ploughed before November to get all the winter snow and frosts; extensive sand and gravel pits; and peat into which a tractor can in places sink almost out of sight (and which is worked for horticultural purposes by the Peat Moss Litter Co.).

Some of the contrasts of Yorkshire farming may be seen if you travel by road or by rail from Darlington to Doncaster. The A1 goes through the better agricultural land, keeping more or less to the magnesian limestone ridge, though towards Doncaster some poorer land can be seen. The main line railway runs east of this ridge, where the quality of the land varies greatly with the drainage. Much of the poorer land can be seen, including more old permanent pasture of the "rigg and fur" type than one might think still existed, and some bits of ancient swamp and woodland. South of Doncaster seems for some miles to be agricultural desolation. (The "Rigg and Fur" or ridge and furrow pastures were made in past centuries by ploughing in small "lands"—just eight or nine times with a horse plough round each main ridge, giving a domed slope of seven or eight feet across, with a channel between. This was mainly for surface drainage and partly perhaps because it gave a larger surface area for grazing. This is no good for modern implements and most have long since been ploughed out and levelled. But some fields of really good permanent pasture remain, and a number that are so poor they are not worth ploughing.)

One last aspect of the geological background of Yorkshire farming is the place-names it has given us. "Carr" was—and often still is—a low-lying, often ill-drained, place of heavy land. In the south of Yorkshire it may also apply to peaty land on top of alluvial deposits, but it is always wet and difficult to drain. The same word carr, from a slightly different Norse origin, means a small wood, usually of alders, on boggy ground. Apart from field names and farm names (Potto Carr, Morley Carr), there are village names like Carthorpe. "Holme" is a meadow near a stream and generally refers to what is now good grassland. "Ing" (more often in the plural) is also a low wet pasture, which may after draining now be arable but usually gives some trouble in wet years. "Swang" is also a hollow wet place, frequent on or near the moors, while "moss" and "syke" refer to similar situations. "Grain" is a fork in a stream, or a side valley. "Mires" and "Marishes" may now be well-drained and productive but throughout the Vales of York and Pickering indicate where the old marshes or lake bottoms lasted longest. "Eller" (alder) and "birk" (birch) may indicate bits of marshy woodland, which have long since disappeared. "Shaw" is also a wood, while "thwaite" is a clearing. "Ley" may be a clearing or shelter, or it may be a grass field. Sown grass, as distinct from old permanent pasture, is still called a ley. "Lund" or "Lownd" is a sheltered farm or field. On any one-inch map a name or two will refer to rock or scar, sand or clay, or some geological feature.

Before leaving this geological chapter, a word should be said on a kindred subject, climate. Though a Yorkshire farmer going to Smithfield Show will swear that London is at least one coat warmer than York, if he goes to visit a cousin at the head of Swaledale or Teesdale he will certainly find it two coats colder there! In Cleveland we read of Hampshire or Berkshire farmers starting to sow, or to harvest, a fortnight or three weeks before we can, but this is often more a matter of soil than climate. The early soils of Yorkshire, the sands and gravels, the "golden mile" near Ripon, and the magnesian limestone, are never far behind the south. York is colder than Oxford in winter, but during the growing season they are the same average temperature. The main difference is that York reaches the right temperature for growth about a week later than Oxford, and falls back again to it a week earlier. The Dales however are a week later still. York gets 250 hours less sunshine than Oxford, and the sunshine is slightly less intensive. But areas near the West Riding towns get far less than York.

Rainfall varies a great deal: The Vale of York gets 24 inches, slightly lower than is satisfactory for growth, while the Pennines get over 60 inches—some western slopes over 70 inches. Because of uplands on both sides, the central Vale of York is drier than anywhere else except East Anglia, and many farmers have gone in for irrigation. The Cleveland Plain, nearer the sea, gets 26 to 30 inches—over 30 inches on the hills. But it suffers a good deal from "rokes," the sea mist which blows in and can stop hay-making or harvesting for days though not depositing much actual water. In the West Riding the number of days on which snow lies varies from an average ten in the Ouse valley to over 50 on the Pennines. In the East Riding it varies between 17 for the lowlands and 25 or more for some of the higher Wolds. In the Pennines climate is an extra factor adversely affecting farming conditions, another reason for the historical dependence of Dales farmers on buying hay and straw from the Vale and the East Riding, selling their store cattle there, and often wintering their sheep on the lowlands. It has often been said that any Yorkshireman going south to farm is bound to thrive— and it usually seems to be so. Equally, after a journey through southern counties, it always used to be a pleasure to get back to Yorkshire and see neat and well-thatched hay and corn stacks again. Even in these days of combine and baler, you will see neater bale stacks in Yorkshire than in East Anglia and the south.

Over the centuries the slightly harder conditions of life have made Yorkshiremen that bit more careful in craftsmanship, and their financial thriftiness is proverbial. Old Benny Thompson from Sowerby under Cotcliffe was hard hit like other farmers in the agricultural depression of the 1930s. If he wanted a hammer or a hay fork he would go to each of the four Northallerton ironmongers in turn to find the cheapest. The lads at the shop he always went to first found this out and started having a game with him. They named a very low price, and after he had been round all the others and returned to buy this unusually cheap tool, they would express sorrow that it had been sold and only more expensive ones remained. Three times Benny went away shaking his head in disappointment while the lads laughed behind his back. Next market day he came in for something rather more expensive—a wheelbarrow for the cow shed. Still full of their joke the lads in charge named a price about £1 less than normal. "Ah'll tak it," said Benny quietly, and as they looked at him in consternation added, "Ah went t'other way roond this tahm!"

2.

The History of Yorkshire Farming

MESOLITHIC men of the Azilian and Maglemose cultures were the first settlers in Yorkshire after the ice receded—about 8,000 B.C. The former came from the south and France, and their remains have been found in the dales. The latter, who had habitation sites at the eastern end of the Vale of Pickering, were the first of our Scandinavian visitors. These were closely followed by a similar race of nomadic hunters who left tiny flint arrow heads—"pygmy-flints"—scattered in large numbers over the Pennine Fells and the North York Moors. It was not until the climate further improved, and the first Neolithic tribes—armed with stone axes—arrived at sometime between 6,000 B.C. and the beginning of the Bronze Age round 2,000 B.C. that primitive farming began to develop. Certainly the Neolithic people—who inter-married with their predecessors—kept domestic animals, cultivated wild grains, and made cooking pots. They had dogs, and in Elbolton Cave, Wharfedale, the bones of sheep, cattle and goats, needles and tools of bone, and a bone whistle, were found. One wonders if here lived an early sheep-dog trainer!

By the Early Bronze Age, properly woven woollen cloth was in use, and new invaders from the Rhine, with better weapons, probably formed a fairly wealthy aristocracy. Stone axes were perforated for a shaft, and soon the wealthier people had metal tools and weapons. Two wheeled and even four wheeled horse drawn carts came into use. There were settled farms now, with sufficient cultivation to provide some bread or havercake and oatmeal porridge. One such site is at Wayworth near Commondale in Cleveland and one at Dewbottoms between Arncliffe and Malham Moor. Improvements in weapons, and presumably in tools, characterised each fresh wave of immigrant tribes from the continent. Iron was certainly known and used in the Late Bronze Age, but the "Iron Age" as such reached Yorkshire about 100 B.C. only just ahead of the Romans. By then agriculture in settled regions must have been fairly advanced. Even on the moors there are many traces of sunken cattle roads, walled enclosures and small fields. In the Dales, Celtic lynchets are numerous (Malham West field; Thorpe West field, Wharfedale).

It is probable that many of the defensive trench systems on the North York Moors, with cattle enclosures included, date from these pre-Roman times when the earlier Bronze Age people were defending themselves on outlying farms against the later Celtic invaders. They were probably still there during Roman times when the Celts or Britons settled down to profitable farming for their Roman overlords. The last twenty years have swept away more relics than the previous two thousand, and before the war it was not difficult to believe that a few of the families in the Cleveland moors and dales were Bronze Age survivals, who had withstood alike the Celtic, Roman, Saxon, Norse and Norman invasions. Since the war, even these have succumbed at last to tractors, electricity and television.

The Romans were good farmers and farm managers. They came to Britain for its grain, and perhaps for its cattle, and they were here for four centuries—as long a time as has passed since Elizabeth I. By the time they left, farming had reached a stage probably more advanced than was seen again until the monasteries took a hand. There were large Romano–British villas, with well organised farms, in many places, but particularly in the Malton–Helmsley area, round the Howardian Hills, at Gargrave and Middleham, and in all the more favourable parts of the Vale and lower Dales. In several villas, grain drying kilns have been found—e.g. Cold Cam near Oldstead—a surprising link with modern agriculture. It is safe to assume that some farms in the upper dales were not so advanced in agriculture and standard of living in 1935 as the villa farms of 335.

What most held Yorkshire farming back after 400 A.D. was the precariousness of it—raids by Scots, Anglo-Saxons, Danes and Vikings continually brought to an abrupt end settled farms and families. There was increasing dependence on the agricultural community (the Anglian "manor" with its three arable fields). We know that some British communities survived through it all—probably our first Yorkshire poet, Caedman, was of British descent, with hereditary skill as a stockman, working as such for Whitby Abbey. Much of our Yorkshire folk lore and fairy stories probably refer back to the British or to remnants of still older races who lingered on in isolated places on the moors—stories of hobs, wise men, fairy cows and the like.

Yorkshire suffered still more severely from the Normans' Harrying of the North. This must have put back agricultural development by at least a century. Seventeen years after the Harrying, in the Domesday survey, the entry "hoc est wasta" appears with painful frequency. In the Honour of Richmond, 108 manors out of 193 were still complete waste. Wensleydale west of Aysgarth was a howling wilderness where wolves, deer and wild boars went unmolested by man. In the Vale of York, Burneston and Pickhill each had only one plough where there had been thirty, and Northallerton was waste. In Cleveland's 120,000 acres there were only eleven farmers (socmen) and 225 villeins and labourers working 58 ploughs.

The Cistercian monks did more than anyone else towards the redevelopment and further improvement of Yorkshire farming. Their monasteries at Rievaulx, Fountains, Sawley, Roche, Meaux, Kirkstall, Jervaulx, and Byland were centres of good farm management and breeding of better cattle and sheep. The monks of Rievaulx turned Bilsdale from a marshy wilderness into good farmland. Fountains had 18,000 sheep at one time on Malham Moor. Jervaulx bred horses and developed cheese-making in Wensleydale. Meaux exported wool to Italy from Holderness. Other monastic orders also contributed, and the "grange" system of farming was wide-spread, as the frequency of that place name suggests. The Cellarer of an abbey was responsible for the food supply (as well as the drink) and for the supervision of outlying granges, where corn and wool were stored until sold. Some of them were better farmers than they were monks, and in 1275 a complaint was made that the Cellarer of Newburgh "trafficks in horses and hath a rough tongue."

Yorkshire farming at this time presents no simple picture. In Anglican villages a manorial three-field system (wheat, barley or oats, and fallow) shared out in strips, with shares in the meadow and common grazing on the waste, was no doubt still predominant, though many manors had disappeared after the Harrying of the North. The monasteries had certainly taken over a great deal of land and were farming it on their grange system. The Norse settlement had been mainly in scattered farms, with sheep husbandry as the chief occupation. There is no doubt that in the Cleveland Hills and Dales, and in the upper Pennine Dales, many of these survived and still survive. The place names are a Viking roll-call—Thorogill, Thorgill, Sarthwaite, Stang, Swang, Breck, Brenk, Broxa, Blakey, Hamer, and Ankness. Some of the Norse farmers had had summer grazing shielings on high pastures—Summerside, Appersett, Burtersett, Raven Seat (from Norse *saetr*). Some of these the monasteries took over and made into big sheep farms. But even some individual farms not so high in the dales still bear names that are recorded as early as 1300—William Beck in Bilsdale and Sleddale Cote near Guisborough.

The Black Death (in the North Riding, May to August 1349) further weakened the manorial system, speeded enclosures, and helped towards transforming villeins into tenant farmers. Nevertheless, Yarm had a three field system till 1658, and in some villages it must have persisted much longer. The Manorial Court survives at Danby and elsewhere. But the three field system was probably never universal in Yorkshire. There are certainly plenty of farm-houses on their own land dating back to the 17th century so that by the time of Arthur Young and William Marshall—around 1780—Yorkshire farming must have been organised on much the same basis as we know today, though there were still some large commons in the

Vale of York, and much of the Wolds and some moorland still remained to be enclosed. Vast areas of moor and fell never have been enclosed.

Our earliest noteworthy record of modern Yorkshire farming is Henry Best's Farming Book, 1641. He farmed at Elmswell, near Driffield, an estate which had belonged to one of the York monasteries, reverted to the Crown, and was bought by the Best family in 1598 (for £2,200). They remained in possession till 1844, when they sold Elsmwell for £42,500. In Henry Best's time, apart from the demesne lands, five main farms were rented off—West Hall, "sixeteene oxegange of arrable lande, besides inclosure"; Laborne, 8 oxgangs "with pasture and meadowes thereto beglonging"; Skelton 8 oxgangs and pasture; Lynsley 6; and West House 8. (The oxgang has been reckoned between 12 and 24 acres. At Driffield enclosure the oxgang was reckoned at 24, but this probably included the pasture and meadow.) If the arable was half the land these farms would be quite large—from 140 to 400 acres. The rent was £2 per oxgang or about 2s. per acre overall (including pasture). Taking the capital increase in the value of the estate as a measure of rental values this would be equivalent to £2 per acre in 1844, which is probably not very wide of the mark. Obviously the Three Field System was in transition, and the pasture had already been enclosed. But Henry Best gives much detail of the numbers of "lands" and "wandills" (a strip worked in one day) and of exchanges that had been made between farms.

There is much information about sheep and shepherding, of harvesting and thatching, of bee-keeping, of marketing:—

> "The ffirst ffayre hereabouts is Little Driffield on Easter Munday; on St. Hellen day 3rd May there is a ffayre at Weeton—there is also a faire at Brands–Burton in Holdernesse; att these three fayres handsome leane beasts, leane weathers, old ewes, and the most timely sorte of lambes have very good vente, because that Holdernesse-men come in and buy up such for stockinge of theire feedinge-grownds; fatte horses, and especially geldinges, goe also well. On Trinity Munday there is a faire att South Cave, att which are many sheepe bought and solde; horses alsoe goe well there, and especially mares, because it is near to Walling fenne, the greate common; and if a mare chance to fall lame, they can put her to the common and breede off her."

There are careful instructions for "Keepinge Waines and Coupes from Wette" by putting them away, with barrows and "stees" (ladders) in "helms" (sheds) for the winter. "Our folkes weare this yeare imployed aboute this business on Powder treason day." There is much information about hiring servants (at the Martinmas hirings, or by keeping good servants on). "Wee give usually to a foreman five markes per annum and perhaps 2s. or halfe a crowne to a godspenny, if hee bee such an one as can sowe, mowe, stacke pease, goe well with fower horse, and hayth been used to markettinge and the like." (A mark was about 62p, the godspenny an outright gift at the striking of the bargain.) "Wee give usually seaven nobles to a third man, that is a good mower, and a good fower horse man, and one that can go heppenly with a waine, and lye on a loade of corne handsomely." (Noble—about 35p.) "Wee give usually to a spaught (youth) for holdinge of the oxe plough 30s. if he be a wigger and heppen youth (strong and active) and 20s. to a good stubble boy for drivinge of the oxe plough and that can carry a mette or three bushell pease out of the barne into the garner." In addition there was much seasonal employment (for hay-time and harvest) of "moorefolk" for whom straw beds were prepared in an out-building.

Obviously many of the labourers both casual and permanent came from the Cleveland moors, and were tough, hard workers, rather superstitious. Other ages noted were:

> Threshers 6d. a day, mowers 10d., boys 2d. or 3d. Mole catchers have usually 12d. a dozen for old moles, and 6d. a dozen for younge ones. Thatchers have in most places 6d. a day and theire meate but we never give them above 4d.—because their dyett is not as in other places; for they are to have three meales a day, viz; theire breakfast att eight of the clocke, theire dinner about twelve, and theire supper aboute seaven or when they leave worke, and att each meale fower services, viz; butter, milke, cheese and either egges, pyes, or bacon. (These day wages were not greatly different from those laid down by the Justices of the Peace for Kingston upon Hull in 1570—"The Mower by day, with meate and drinke 4d., without meate and drinke 10d. The thatcher —4d. and 8d. Bayliffes of husbandry by yere, with meate and drinke 25s.; his livery 6s. 8d., Shepherdes, with meate and drinke, 21s., his livery 5s., one quarter of pease, one quarter of barley." The wages of the more regular servants had however almost doubled.)

A century and a quarter after Henry Best we have detailed information from that early agricultural economist Arthur Young, who was to be the first Secretary of a new Board of Agriculture. His *Six Months' Tour through the North of England* was published in 1770. He was not impressed by the Vale of York, a lot of which had not yet been enclosed, or enclosed and not yet improved by the essential drainge. He was much more impressed by Cleveland, already well enclosed. He found Holderness more than half grass. Draining was in progress and improved the letting value from 5s. an acre to as much as 35s. Eight years after Young we have even greater detail from a Yorkshireman, William Marshall, whose *Rural Economy of Yorkshire* (1778) is fascinating reading. It is astonishing how modern Marshall is in his enthusiasm for new species of grasses in leys—particularly ryegrass—for liming and for other things which might still be preached by an agricultural adviser. And, like a good Yorkshireman, he includes one of the earliest dialect glossaries to give the foreigner some chance of finding his way around. To deal with Marshall in any detail would fill the rest of this book. No student of Yorkshire or of farming should miss reading his two volumes. Here are some examples of his trenchant style.

Turnips and clover leys were both spreading rapidly. Some

farmers were still fallowing for wheat because they believed that to sow it after clover bred quicks (couch). "There is no worse management than sowing wheat on a *foul* clover ley; but this is no argument against ANNUAL LEYS. If the land be clean when the clover seed is sown it will as soon breed sugar canes as quicks." He speaks enthusiastically of the many horses bred in the Vale of Pickering ("between five and ten thousand, between the Eastern Morelands and the Humber"), but is very scornful about the "Fen Breed, the Howden mack of Black Cart Horses, which I am sorry to see worming their way into the Vale. The breed of grey rats, with which this Island has of late years been overrun, are not a greater pest in it, than the breed of black fen horses: at least whilst the flesh of horses continues to be rejected as human food. They are a breed better calculated for eating than working, and render their drivers as sluggish as themselves."

Seventy years previously, an ancient breed of black cattle were the only ones seen, but these were succeeded by the Long horned Craven breed, then by a short-horned Holderness breed of Dutch extraction—"probably the worst breed the Vale ever knew". But now a much-improved Tees-water shorthorn—"a most valuable breed" was coming in. Marshall goes into some details about horns: "I am not a BIGOT to horns of any shape or length. I would as soon judge of a man's heart by the length of his fingers as of the value of a bullock by the length of his horns. If his flesh be good— I would much rather have him without horns, than with any which enthusiasm can point out. The doctrine of horns has long appeared to me as a species of superstition among farmers, and as a craft convenient to leading breeders." His judgment of cattle is to some extent affected by the fact that oxen were still in use for haulage, though "two horses, with whip reins, without a driver, is now the universal plow team for all soils" (previously, four or six oxen, led by two horses, had been common "And, in breaking up a fallow, two men and a boy were the common attendants of this unwieldy expensive team"). A much improved plough had come into use, and improved, hard, "barrelled" roads, bad for the feet of oxen, were speeding their decline even for haulage. Marshall regretted this, since the ox was still valuable for meat after its working life was over. "An ox which I worked several years might, at seventeen or eighteen years old, have challenged, for strength, agility and sagacity, the best bred cart horse in the kingdom."

Marshall confines his attentions to the eastern half of Yorkshire, and particularly his native Vale of Pickering. He describes the "Western Morelands" (the name of "Pennines" had only just been invented by Charles Bertram) and the fertile district of Craven, but considers "the Western Division unfit as a subject of study." Though there were no doubt highly cultivated lands, especially about Doncaster and towards Ferrybridge, yet "no country which is entirely mountainous, nor one which is disturbed by manufacture, can be a fit subject of study for rural knowledge." Apart from its natural advantages, the East had many industrious farmers, "while the SPIRIT OF IMPROVEMENT, which has lately diffused itself amongst all ranks of men, renders this district singularly eligible as a field on which to trace the greater outlines of management."

In doing this, apart from subjects already mentioned, Marshall deals with tenancies, commons and enclosures, farm buildings (passing scornfully over "fancy farm houses"), hedges and fences, the planting of new woodlands, and selection of trees, soils, manures and weeds. He is perturbed by the paper money rapidly coming into circulation from private bankers, and the inflation that had resulted (which had enabled some rash farmers to pay off debts more easily). "The price of manufactures depends upon those of materials and labour; and this on the PRICE OF LIVING. If by a flow of cash in circulation these prices be suffered to advance, the demand for manufactures will of course decline, and with it the prosperity of the nation." (This is 1797, not 1972!). Apart from consideration in great detail of the usual crops, improved varieties, and methods of sowing and harvesting, he has a fascinating chapter on flax, for which there was then a good local market in coarse linen mills in several places in Cleveland and the Vale of Pickering. Surprisingly, he also deals with tobacco, of which a good deal had been grown about 1782 "cured and manufactured by a man formerly employed upon the tobacco plantations of America". However, the larger growers had run foul of the law!

William Marshall was a popular but practical writer who certainly contributed a great deal to the further improvement of Yorkshire farming in the years that followed. In 1794 we get the Board of Agriculture Reports—the *General View of the Agriculture of the North Riding* by J. Tuke, of the *East Riding* by I. Leatham, and of the *West Riding* by George Rennie, Robert Broun and John Shirreff. These are more balanced and comprehensive than Marshall, but nothing like such compulsive reading. In the North Riding average rents were then 15s. to 21s. an acre—as high as £3 or £4 near large towns. Flax had declined, turnips increased, and turnip drills were being used. Teeswater shorthorns were the general breed. Except in Cleveland, a large proportion of farm-houses were still disadvantageously situated in villages. Wages were from 1s. to 2s. a day, without food, having apparently doubled since Henry Best's day—though certain piece-work rates were at least three times as high—threshing wheat 2s. a quarter as against 8d., oats 1s. as against 4d. A yearly wage of £16 for a head man compares with £3 or £3 10s. in Best's day, though with rising prices it is doubtful if men were much better off. Labourers generally were very badly housed in low, damp, one-room hovels. Some of the better ones of that time still survive, usually two-roomed, long and low, and looking quite attractive now with improved interiors and mellow brick. At this time wheat was round 6s. a bushel, oats 3s., barley 4s., cheese 3s. 3d. a stone, and bacon 6s. 3d. a stone. Tuke praises Yorkshire horses and ploughing—"A Yorkshire ploughman will by himself plough full as much land in the same time with two horses, and full as well, as a south-countryman will with four and a driver. In favourable circumstances two acres a day are frequently ploughed." There were a few open arable fields left in the North Riding, but not many.

For the East Riding, Le atham adds nothing to Marshall, but comments that the township of Hunmanby was still mostly open fields and commons, but managed on a more modern system with everyone agreed on improved methods of farming. For the West Riding the three writers were clearly intent on proving Marshall's opinion wrong. "Manufactures have had a sensible effect in promoting agriculture in this district" by offering a ready market. They may have raised the rate of wages but they have also raised prices. This in fact was "by far the most valuable of the three Ridings, whether considered by magnitude, fertility of soil, or population." The Report comments on land near manufacturing towns, "occupied by persons who do not consider farming as a business but as a convenience. The manufacturer has his inclosure, wherein he keeps milk cows for supporting his family, and horses for carrying his goods to market, and bringing back raw materials." The magnesian limestone strip was already good arable. Round Boroughbridge and Wetherby half the fields were under the plough, further south two thirds. There was not much waste in this section, but still too many common fields "crying out for immediate improvement . . . Generally of the best soil, they carry the worst crops . . . Land rises in value the moment it is divided and free scope allowed to the genius and talents of the farmer." Again, too many farms were still in villages and not on their land, and there was a great dearth of housing of any sort for labourers not living in. The fact that the burden of maintaining roads by statute labour (six days' labour of a horse team and two men for every £50 rent) fell unfairly on the farmers alone was particularly noted in the industrial West Riding. The district suffered particularly from the smallness of farms and the absence of long leases; uncertainty of tenure was the greatest obstacle to progress. "Before a farm can be put in proper order a considerable time must elapse, and much money must be expended. Near twenty years may take place before a tenant receive full reward for improved cultivations."

Everywhere in Yorkshire improvement went on steadily during the 19th century. There were further official reports—Strickland in 1812 and Howard in 1835 on the East Riding. Cobbett visited Yorkshire in 1830 and Rider Haggard in 1906. There were Royal Agricultural Society Reports in 1848 (Legard) and 1861 (Wright). Generally, after good prices during the Napoleonic Wars there was depression for many years afterwards, with low corn prices, and an unusual number of wet seasons, leading to low yields and disease in stock. There was a great deal of emigration to America and Canada at this time, particularly from dales and moorland farms. On an 1841 map of Upper Canada, describing new settlements from Quebec to Lake Huron, can be found (north of Toronto) "East Riding", which in turn includes "Pickering," "Whitby" and "Scarborough" (still on modern maps). There is also a "Flamboro," and "Burlington Bay" (the old name for Bridlington) at the head of Lake Ontario. There is also a "West Riding" and, south-east of Buffalo on the modern map, a "Yorkshire." Letters home from several emigrants from the Danby–Glaisdale area throw a revealing light on conditions in Yorkshire round 1830. John and Ann Knaggs write, from Nelson, Ontario, October 1830:—

> In haytime a man has 6 York shillings, in harvest 8 York shillings, a day— there is plenty of work and good wages, not tossed about from house to house for 8d. a day as in England. If all poor labouring men was here it would be much better for their families ... I often wishes you with some of this fine light bread. We never saw a bit of brown bread since we came ... It is a pity there are so many privileged in America and so many distresst in England ... I can live as well as some of the gentlemen of your place for I can have beef or pork or fresh meat three times a day and Tea three times a day.

Conditions in England in fact were so bad that a Parliamentary Select Committee on Agricultural Distress was set up and its report (1836) emphasised the depth of the depression, particularly in Howden and Holderness, where rents had had to be reduced by a third. After 1850 things began to improve and the third quarter of the 19th century was the "golden age" of farming. Industrial prosperity and urban populations were increasing rapidly, and to help meet a rising market new agricultural machinery was being invented, seed varieties being improved and selective breeding practised. The upland areas benefited from rising prices—the margin of reclamation and cultivation probably went higher than it has ever been since— but not so much from the scientific advances. Long after the self-binder—or even the early combines—had been introduced on the plain, you might tie sheaves of corn by hand behind a "manual" reaper (I did this in Nidderdale as a boy in 1922) or turn hay by hand-rake (well into the 1930s). The flail was still used on occasion for threshing small lots of corn in Glaisdale and in Snilesworth in the 1930s—perhaps later.

There was depression again at the end of the 19th century and another Select Committee on Agricultural Distress. War brought prosperity, followed by perhaps the worst depression of all in the 1920s and '30s. There was complete neglect of farming by Government, high imports and low prices. For most farmers life was a struggle for survival and many failed to survive. Some farms could not find a tenant when offered rent-free, and some large farms in

Holderness were sold for £7 an acre. They would probably fetch £200 or more now. Those who were able to hang on reaped high rewards during and immediately after the second world war.

Whatever changes in farming had gone before, were nothing compared with changes after the war. Since 1950 a new agricultural revolution has been going on at an ever faster pace, and this time the uplands have moved as fast as the lowlands. The ghost of any Cleveland farmer of 1750, returning to the Cleveland of 1935, would not have felt greatly out of place. In the Cleveland of 1972 he would be lost completely. The same might be said of almost anywhere in Yorkshire, though on the largest and most advanced arable farms of the Vale of York and the East Riding the ghost might have already been a bit shaken in 1935. One factor since 1950 is that, quite apart from the technical changes in farming which include, as compared with 1935, the complete disappearance of the horse as a working animal, any old village within twenty miles of a town has doubled in size, and has a preponderately non-farming population. Farmers are now outnumbered in their own villages, where they share the pub and the parish council with business executives from the towns. The farmers themselves are business men, accustomed to talking of gross margins and cash flow, and applying sophisticated systems of economic analysis to their farming enterprises. A modern 250 acre farm with an 80 or 90-cow herd may represent a capital of £100,000. To survive in the 1970s with a reasonable standard of living requires ruthless efficiency, the minimum of labour, full mechanisation, specialisation in one or two enterprises, and careful accounting. Even now there are a few Bronze Age survivals here and there, content with a subsistence level of farming, unwilling or unable to change, but they grow rapidly fewer. Between 1944 and 1962 the number of agricultural holdings of over five acres in Yorkshire dropped from 29,574 to 25,721—over 200 farmers leaving the industry every year—and the rate has been even faster since then. In the same period the number of agricultural workers dropped from 58,363 to 43,050. In the process much of the old pleasure and comradeship has gone out of farming. There are no longer co-operative threshing days, with a dozen neighbours helping, and everything on the farm is a one man job. The threat to the continuance of the family farm comes now not from Scottish raiders, but from Death Duties, Capital Gains Tax, inflation, and the impact of big business in the form of integrated production and supermarkets—deadlier enemies than the Black Douglas. To survive the next twenty years Yorkshire farmers will need all the tenacity they have shown over the past two thousand.

3.   *Cleveland*

WHEN Arthur Young rode over the Hambleton Drove Road into Cleveland in 1769 he was appalled—agricultural improver as he was—by the "vast waste of desolate moors," but greatly impressed by the view from Scarth Nick over the Cleveland Plain, looking across Swainby and Potto towards what is now Middlesbrough but was then just two farm houses on the mud-flats by the Tees. He described it as "an immense plain, finely cultivated, and the in-closures adding prodigiously to the view." Thomas and Emmanuel Bowen's Pictorial Map of North Yorkshire in 1777 has the note "Cleveland or Cliveland in the Clay, so called from the high rocks and precipieces with which the parts abound and Soil being of an exceeding clammy stiff Clay. Here is made very good cheese, not inferior to that of Gloucestershire." Strangely, the Bowen map makes no mention of Wensleydale Cheese and it is probable that in the 18th century Cleveland cheese was more important. The third day of Yarm Fair every October was traditionally the cheese day and in 1820, 683 carts came into Yarm for this fair each carrying 25 cwts. of cheese. Only a few old people remember Cleveland cheese, though four large blocks of dressed sandstone from a cheese press still stand in the stackyard at Potto Hill. The growth of Teesside in the 19th century, and the coming of the railway—the Picton–Potto–Stokesley branch was opened in 1857—led to a rapidly increasing demand for liquid milk and a more profitable basis for Cleveland agriculture.

Tuke commented in 1794 that "wheat is the staple produce of Cleveland. No other district in the Riding, or perhaps in the North of England, produces as much in proportion to its size, or of as good a quality." The heavy boulder clay of Cleveland still produces good wheat, though as in other regions there is more barley and less oats than in former years. On the Cleveland soils a cold wet spring can be disastrous and there was a series of them from 1967 to 1970. Some fields cropped profitably in the 1950s have had to go back to grass. Wheat, sown in the autumn, tends to be very much the most profitable corn crop in these conditions. It is very rare indeed for cereals on the Cleveland clays to suffer from drought. Partly because of the threat posed by wet seasons, and the existence of a fair amount of land where drainage is a problem, Cleveland is very much an area of mixed farming. The mainly arable farm is a rarity except on the coastal plain near Redcar and one or two parts near the western boundary of Cleveland towards the Tees. Grass remains an important crop and the really typical Cleveland farm is half milk and half corn, though many farmers substitute beef nowadays for the milk. Some have a large pig or poultry unit as the main live-stock enterprise, and a few retain a flock of lowland sheep.

Only a couple of miles and a thousand feet separate some of us on the Cleveland Plain from the entirely different conditions—and climate—of the Cleveland Hills and Moors. Many farmers on the plain are of moorland stock, as successful hill farmers have moved down or sent their sons down onto the plain, over the generations. "We've sent more than store cattle down," said one of them, "We've sent store men as well!" One such family were the Fells. Among letters from emigrants to Canada was one from John Fell, Cavan, Ontario, 1859. "I would not come back to rent the best farm Lord de Lisle owns," he writes (Lord de Lisle was landowner in the Ingleby Greenhowe area). After renting a farm, John Fell first had acquired 600 acres "of wild land" which he was rapidly improving. "Here you are not hampered by Lord nor Lady, nor any of their underlings—all men are equal here." His nephew, James Fell, is listed in the 1890 Directory as "Yeoman farmer"—that is owner of Borough Greens Farm, Easby, just under Captain Cook's Monu-ment on Easby Moor. When they were only 13 or 14 years old, James sent his son (J.W.) and daughter (Ethel Mary) to look after another farm of his in Baysdale, a lovely isolated valley lost in the moors. With this experience behind him the young J.W. took a farm on his own at Ingleby Barwick and later moved to the other side of the Tees at Portrack, an area of flat riverside land now given over to industry. This would be in the 1890s. He prospered, and now his sons J. R. and J. W. Fell farm getting on for a thousand acres be-tween them on each side of the Tees near Croft. J. R. (John) Fell has been a pioneer in the use of irrigation, intensive grass and grass drying. His son David now farms 3,000 acres in Western Australia.

My own farm, Goulton Grange, may be taken as fairly typical of the Cleveland Plain. It is just over 200 acres of mainly boulder clay, with here and there towards the southern boundary and the hills a morainic mound of sandier land. (Here the small farm of Potto Hill, 50 acres, was added in 1958.) The name comes from the family of Goulton or Gautun of Norman origin, who are men-tioned in early records (around 1300) as companions of the Baron Nicholas de Meynell of Whorlton Castle (a few fields away on a hillock above Swainby village). The local pronunciation of the name is still "Gowton" (strangely enough one of my great-grandfathers was Frank Meynell, blacksmith at Crathorne and Picton). In the latter half of the 19th century, Goulton Grange was part of the Potto Hall estate (Joseph Richardson Esq., J.P.) and James Pybus was tenant. In his time a new house was built—a large, three storeyed building, the bricks for which came on the Swainby–Scugdale branch line from Potto station and were dumped in the Greens—the one field that lies between the line and the house. This branch line was built in 1857 to take iron ore out of Scugdale, and was derelict by 1900. The old house still stands, a workshop now, built with narrow hand-made brick about 1700.

The "old line" is now nothing but a strip of neglected woodland running along the Goulton Beck. Just after the 1914–1918 war there was unemployment among returned soldiers in the area and the N.E. Railway employed them to clear the line of scrub. Money gave out when they were half way along the Goulton Grange boundary and this point can still be seen—all is now overgrown again but from there on the growth is much stronger! About this time the Patersons were tenants at Goulton Grange—they moved on later to farm in the Boroughbridge area. At this time, too, Strickland Walkington came down to the Cleveland plain from Beadlam Rigg above Helmsley. He managed Busby House Farm for Major Gloag for some years and then took the tenancy of Goulton Grange in the early 1930s. On the break-up of the Richardson estate in 1950 he bought the farm as sitting tenant for around £9,000 and a year later sold it to W. Cowley & Partners for just over £14,000—indicative of the value of secure tenure conferred by the Agricultural Acts of 1947 and 1948 (a farm with vacant possession can sell for up to twice as much as it would if subject to tenancy). The same farm, with improvements, is in 1972 worth £50,000 or more—a measure of the general inflation, and increase in land values, in the past twenty years.

Walkingtons first bought a tractor in 1934—until then all work was done with horses and some horses were kept till 1957. There were generally 60 acres in corn, about the same amount of permanent grass, and 80 acres of leys that had been down from one year to five years or more. The permanent grass, including the 20 acre Moor Field which had not been ploughed in living memory, was in "rigg and fur." There were 30 milking cows, a flock of 60 breeding ewes, a few hens and a few pigs. This was the typical "mixed farm" of that era, when farmers—used to the hardships of the depression—feared to have all their eggs in one basket. It was a good system of farming, with each department helping the others—when milk was down, pigs were up, or when pigs were down sheep helped to redress the balance. The snags were that labour was dissipated among several enterprises, scattered in various buildings, and capital might also be wastefully employed. An astonishing thing was that a farm of this size had no electricity and we did not get it till 1954. It is impossible to run a modern farm without electricity and there are now at least

14 electric motors in use apart from domestic appliances.

In 1951 we began in a similar way at Goulton Grange though we tried to intensify—the hens became a 1,000 bird laying unit, the pigs a 12-sow breeding unit. We only had 15 cows to begin with and it would take time (or more capital) to expand in milk. We had a 30–40 ewe flock. But the 1950s were a decade of specialisation and ever-increasing necessity for economic efficiency. First the sheep, then the pigs, then the poultry were dropped and by 1960 Goulton Grange—like many other Cleveland farms—concentrated on two enterprises only, cows and corn. A cow-yard and milking parlour had been built, for 60 cows. All the permanent pasture had been ploughed out except for four acres of broken banks by the beck. The "Moor" (after a century or more in grass) had yielded an excellent wheat crop, followed by good barley for a year or two. A trailer combine had been purchased—our last communal threshing days were in May 1954—and also an in-sack corn drier (with oil heater and electric blower) and four aluminium corn silos holding 20 tons each.

Grass was heavily fertilised for intensive grazing and high yields of silage, and 100 acres of corn were grown. Even in a low rainfall area like the Cleveland Plain (26 inches), hay making is always a risky business, particularly when drizzling days of sea-roke occur. In silage making, grass is cut and moved—either fresh, or after one day's wilting—to the clamp where, stacked and rolled by tractor, and as much air kept out as possible, it ferments in its own juice and makes a sort of pickled grass which retains much more of the original nutrients than does hay, as well as being more succulent. If stacked to settle to about 6 feet high, with a width of face of about 6 inches per cow, then cows can self feed at the clamp at one or both ends, and no manual labour need be involved at any stage. Winter feeding then becomes another form of grazing. An early form of silage making, in concrete towers, had been used by one or two farmers in Cleveland, as elsewhere in Yorkshire, before the war—perhaps even in some wet years in the latter part of the 19th century. The practice had not spread because the machines had not been invented to handle the grass. Goulton Grange was one of the first farms in Cleveland to adopt the new methods, and has had full self-feed silage since 1955. Dairy cows will eat 1 cwt. a day of silage, and seven tons per cow is the usual calculation for a winter's requirements—560 tons of silage for an 80 cow herd.

This general trend towards larger herds, loose housed in shed and yard, and milked through a "parlour" in batches of four, six or eight, in contrast to the older byres where cows were each milked in their own stalls by machines carried round, means a great saving in labour—a reduction in the man hours required per cow per year from over 150 to under 50. The same trend to larger units is even more general with pigs and poultry. A small farm tends to specialise

The Yorkshire farming scene. A view of upper Swaledale from Whitaside showing Dalesbred sheep, traditional stone walls and barns, and valley-bottom meadows. (Geoffrey N. Wright)

Contrasting farmscapes. Opposite, top: North York Moors. Ryedale, looking towards Hawnby, showing a "layercake" of arable land, pasture, wooded slopes and heather moorland. (Bertram Unné). Opposite, bottom: "Industrial" West Riding. Abandoned hill farm above Haworth with land "going back" to the moor. (Bob Collins). Above: The Yorkshire Dales. Steep-sloping meadows near Reeth, Swaledale, with hand-raking still being practised. (Betrram Unné).

Sheep, the mainstay of upland farming. Left: Ewes in the snow near Tan Hill. (Geoffrey N. Wright). Above: Sheep and lambs at Hartlington Raikes, Wharfedale, on their way to summer pastures on the moor. (Bertram Unné).

Harvest patterns. Opposite: Two views of hay-making in Wensleydale, the lower picture showing the pikes of the pre-bailing era and the proximity of operations to the water's edge. (Top, John Edenbrow; bottom, G. H. Hesketh). Above: Combining on the rich, rolling uplands of the Wolds. (John Edenbrow).

Two Sleddales. Above: Isolated farmstead well over the 1,000 ft. contour near the meeting of Great Sleddale and Birkdale Becks in upper Swaledale. (Bertram Unné). Below: Sleddale Farm, a green oasis in a sea of Cleveland heather above Hutton Lowcross, near Guisborough. (Author).

in one branch, a large farm to expand each of its enterprises to the extent that it can employ one full-time specialist cowman, pigman or poultryman. One visible result of this is the number of new farm buildings which have gone up in the last ten years. There has been some protest from planners who should know better and from romantic country lovers who would like to keep farm buildings as a museum piece of the 1890s. This is nonsense. Nothing looks better than a well-farmed stretch of countryside with well-designed and efficient modern buildings with asbestos roofs, well-ventilated, easy to clean, and requiring no maintenance. They are certainly better to look at than some of the antique structures of mellow broken brick and warped red tiles where calves get virus pneumonia, no modern machinery can enter, and which should have been condemned as rural slums half a century ago.

This is not necessarily to approve of some of the more extreme forms of factory farming which have crept even into Cleveland in recent years. One large business organisation has developed a vast complex of asbestos poultry houses on the lower slopes of what was previously one of the most beautiful parts of the Cleveland Hills. How this ever came to be permitted by the planning authorities, on the edge of a National Park, passes the wit of man to understand. But where a small group of new buildings is designed as the working nucleus for the land around, nothing could look more appropriate. Even the factory farms may be considered less objectionable than some of the housing development which has more than doubled the size of many Cleveland villages in the last ten years. Some of the old villages, round their wide greens, were much admired, but there has been no attempt to plan new modern villages which might have been beautiful for the future. Instead, typical suburban additions have been allowed to surround or "in-fill" the existing villages. There have been restrictions of a sort, but no positive planning whatsoever.

One other visual development is indeed general, but more startling in Cleveland than elsewhere, because this was one of the breeding centres, and for long the stronghold, of Shorthorn cattle. Prominent among the great early breeders was Thomas Bates who farmed at Town End, Kirklevington—now a mainly arable farm (He died in 1840 and there is a stained glass window to his memory in the church). Shorthorns were once universal in Cleveland—the Walkingtons' herd at Goulton Grange was shorthorn. Very few of these red, roan or blue-grey animals are seen now. They were supposed to be dual purpose, providing either milk or beef, though a lighter "Dairy Shorthorn" was developed. Their critics said they produced neither milk nor beef sufficiently well. For milk they were first supplanted by Ayrshire herds, but in recent years there has been a very rapid swing to the black and white Friesians which really do produce both milk and beef in large quantities. At Goulton Grange we still have a few Ayrshires left but have bred most of the herd over to Friesian. The same policy was pursued rather more rapidly by some near neighbours, the Hutchinsons, at Raven Hill—a farm very similar in size and management to ours, though after trying silage for a few years (but not self-feeding) they went back to hay-making. Next to the Bates' old farm at Kirklevington, Digby Barker has Hill House and Knowles Farms—about 400 acres. He used to think 30 shorthorns in a byre were a lot of cows but now he too has 80 Friesians. The Barkers came down from the moors—from a small farm at Osmotherley—in the late '20s, going first to Potto Carr and then to Hill House. One of my first memories is travelling (in 1920, at the age of 5) in Charlie Barker's carrier's cart from Trenholme Bar station to Osmotherley up the steep Clack Bank. The fields below Osmotherley were white on fine days with sheets of linen from the old mills laid out to bleach. Now railway line and station have disappeared, and Clack Bank has been ironed out in a dual carriage-way.

One of the great modern breeders of Ayrshires in Cleveland, Jack Brunton, did not come from the Cleveland Hills (though driven away from Teesside by the encroaching town, he has gone back to them by purchasing the Urra Estate in Bilsdale). His great-great-grandfather was from Peeblesshire and was shipwrecked on the Cleveland coast, finding employment as a shepherd with Lord Zetland. His son took over Marton Moor Farm and became known as Tallyho Bob—he always carried a bottle of gin in his pocket when he went hunting! In the fourth generation of the family at Marton Moor, the farm which had given its name to a famous Ayrshire herd and many fine bulls was swallowed up by the encroaching town and the herd dispersed. Jack Brunton still farms 600 acres at Ryehill, a mile further away from the present tidal edge of bricks and mortar, where strangely enough he has gone back to Short-horns, or at least to the Blue-grey Shorthorn-Galloway Cross, which he runs with a Hereford bull as a suckler herd for beef production.

The demand for high butter fat Jersey milk is universal and there are several Jersey herds in Cleveland. Although its native habitat is the Channel Islands, the Jersey is remarkably hardy and can be found in some bleak places. On the hills just above Goulton—at High Farm, Faceby—Keith Beaumont has a well-known Jersey herd. High Farm has special problems shared by many hillside farms in Cleveland which are not actual hill farms. They are marginal in the sense that they are on the fringe of the hills but they do not get the special hill farm subsidies. The problem of drainage is often acute on these slopes where contractors cannot use heavy machinery economically. As one man said, "I can't get a drainage contractor to come and look at my land, let alone afford to pay him!" There is also a quick run-off of rain, and leaching of lime and fertilisers. Costs are much higher than on the plain which may be only half a mile away and 200 feet below. One way of getting a higher return on this land, claims Keith Beaumont, is by keeping Jerseys, which are almost as efficient foragers as hill sheep. Five Jerseys, he says, will live on the same land as three Friesians, and with the premium on Jersey milk will pay better.

Another man with a Jersey herd on difficult land is Joseph Sunley at Girrick. North of the Guisborough–Whitby moor road, Girrick (or Gerrick) is in that area of East Cleveland between the moors and the sea where the main iron-ore workings occurred. Grange farm is four miles from the North Sea and from the highest cliff in England —700 feet. Beyond its southern boundary the heather goes up to almost 1,000 feet on Danby Beacon. Joe Sunley—grey haired and soft-voiced but still hard-working at 66—was a fitter and turner at the Cleveland Steel Works in Guisborough during the depression of the 1930s. He joined the Engineering Union to get better conditions, and got the sack instead. There was no other work available so he swapped a gramophone for a nanny-goat and started farming. He rented a small allotment with a wooden hut and then managed to buy a cow. While he was away a bag of cattle cake was delivered —the cow got at it and burst herself. Joe was back again where he started.

A local tradesman offered to help him. "Thoo's trying, lad," he said, "Dista knaw wheer thoo can finnd another yan?"

"Aye—for £15."

"I'll lend thee it—at six per cent—to be paid back in twelve months." It took thirteen but the lender was satisfied, "Ah thowt thoo'd nut be knocked doon!"

Joe had been courting for years while out of work. Now he got married—and for months had no more than eight shillings a week to give his wife for housekeeping. Success was as much due to her hard work and saving as anything. He got a four acre cow-keeping from the Council, and soon had four cows and a milk round. He had to collect grass from the hedgerows and from gardens and was sometimes so tired that if his scythe point stuck in the ground he hadn't strength left to pull it out. A chance came of getting a farm and he went to see the agent.

"Have you got £1,000 in the bank?" asked that gentleman.

"If I had, I wouldn't want yon spot!" answered Joe. He hadn't even a thousand ha'pennies. Needless to say he didn't get that farm, but in 1939 another chance came with Grange Farm, Girrick, and he took it.

"Not being a farmer," he says modestly, "I was very lucky to get it." The land was tough, heavy boulder clay—"three horse land"— and cold. "Never sow anything at Girrick before the second week in April." Sometimes he couldn't turn his cows out till the second week in June. The soil was deficient in most things and he had to budget carefully to buy every bag of basic slag he could afford. One seven-acre field at the moor edge gave two cart loads of loose hay the first year. In 1969—thirty years later—he took 680 bales of good hay off it. Joe came to Gerrick with six cows—Shorthorn and Jersey—and the day he moved he went round his old milk round early to get a few extra shillings. Soon he got into pure Jerseys and now has 25, plus followers. He aims at as much winter milk as possible and finds the Channel Island premium of 5p. or 6p. per gallon well worth while even on this bleak farm.

Even more important to Joseph Sunley than his Jersey cows are his Cleveland Bay horses. These are the famous local breed, evolved from the "chapman" or carrier's horse, and closely related to the 78 foundation mares of the thoroughbred, most of which were located in North Yorkshire. Marshall praises the Cleveland team of "strong, active, coloured coach horses" as being "the stiffest, the most handy, and for level country and long journeys perhaps the most eligible team that invention is capable of suggesting." The late Sir Alfred Pease in his notes in the Cleveland Bay Stud book mentioned a Cleveland Bay which carried 700 lbs over 60 miles in 24 hours four times a week, another which did the 202 miles from York to London in three days and another which would take its rider 40 miles to Newcastle and back at night, swimming the river Tees on each journey! Joseph Sunley's grandfather had been horseman to Sir Alfred Pease's father and Joseph bought his first Cleveland Bay, a two year old stallion called Lucifer, at Sir Alfred Pease's sale for £10. Joseph worked his farm at Gerrick with Lucifer (and other Cleveland Bays) right through the war, and used him for breeding too, finally selling him for £100 in 1947! "You couldn't do that with a tractor—anyway, I couldn't afford to buy paraffin then, let alone a tractor!" But the man who was "not a farmer" has gained international fame as a horse breeder. From the cold fields of Gerrick, horses have gone for the Queen's carriage, for the Government of Pakistan, and for the Emperor of Japan. The unemployed fitter who started with a nanny-goat has had an award from the Queen's own hands at the Windsor Horse Show. "Shire horses were slugs compared with Clevelands," says Joe. "The Cleveland Bay is a man's horse—it'll go all day."

I am reminded that when we came to Goulton, one neighbour Arthur Bell still had a Cleveland Bay to do most of his work on 40 acres. One day he was ploughing with this horse (a light one horse plough) when the Hunt came by in full cry. The Cleveland was off after them, plough and all, so Arthur cut the traces, jumped on its back, and was in at the kill, getting the fox's brush presented to him!

4. *The North York Moors and Dales*

THE main water-shed of the North York Moors, the route of the Lyke Wake Walk, is 35 miles from east to west as the crow flies, from Beacon Howes above Ravenscar to Scarth Wood Moor above Osmotherley. It lies mainly between the 900 feet and 1,500 feet contours, the greatest height in fact being 1,492 feet on Urra Moor. With the single exception of the short glacial lake valley of Scugdale, which cuts through this watershed towards the north, all streams to the south flow into the Derwent/Humber system, and all streams to the north (with the single exception of the river Wiske) flow into the Tees or the Esk.

For centuries the main—indeed the only—stock of the high moors has been the Black-faced "Scotch" sheep. Marshall noted that in the 18th century these were kept at the rate of one to ten acres of heather —between, 20,000 and 30,000 sheep on the 200,000–300,000 acres of uncultivable moor. He estimated the profit at half a crown per sheep or 3d. per acre a year, and enclosures of moorland made in his day were selling for two or three acres to the £1. This probably underestimated the profitability of moor sheep at that time and certainly in the following century, as Marshall himself (with a footnote about "little fortunes" made in the dales) suspected. Labour was scarcely considered an expense in those days, and the moor sheep cost practically nothing but labour. There might be many losses in bad years, but whatever was sold off the moor was cash income with little cash outgoing. It was the moor stray, and the flock "heafed" to it (native to that moor for generations, and sold with the farm, or at valuation with the tenancy) which made the real value of most dales farms.

Harry Todd was tenant at Holme Farm, just under Clay Bank at the head of Bilsdale for fifty years from 1904. At first he had no moor stray but his sheep would get out over the narrow ridge and down the rough wooded slopes of Ingleby Bank. The agent complained—so Harry offered to rent the woodland—and eventually got it for £6 a year. "Well—£6 for 200 acres of moor-edge and bottry bushes (elders) was nowt. One fat lamb a year about paid for the rent." Harry kept 60 or 70 ewes on that hillside, fattening extremely well through the summer on ash and elder leaves. He sent the lambs off fat from the woods and never brought the sheep into his farm land unless there was a very snowy fortnight in winter. "Yon banks made t' spot," he said, and his successor, who because of re-afforestation has not got the stray, has hard work to make ends meet on the same farm. Of course this woodland would be better grazing than open moor, but it does give some idea of the importance of a stray. And some moor strays are for 150, 200 or 400 sheep.

Mr. M. F. Graham of Hunt House, Goathland, runs a thousand ewes and followers (probably 750 lambs a year) on the 3,000 acres of Wheeldale Moor and the 1,000 acres of Goathland Moor crossed by the Lyke Wake Walk. One ewe to four acres is much closer stocking than Marshall suggested. The 300 acre Scarth Wood Moor (National Trust) has been assessed at 90 ewes, with 25 per cent replacement stock, for the purposes of the Hill Sheep subsidy. This now gives the hill farmer £1.50 per hill ewe and his annual return on a ewe, including this, might be about £3, though for some farmers on difficult moor the subsidy will be all the net profit they make.

The fact is that in 1971 labour has to be considered a real cost in the keeping of moor sheep, and fewer and fewer people are prepared to undertake this arduous life. Half the farms in Bilsdale have now given up their moor strays, and prefer to intensify their dale bottom farming, with cows and milk. So far others have taken over the stray and the numbers of moor sheep have not altered greatly—there are about 45,000 hill ewes and maybe 11,000 or 12,000 gimmer hoggs (the young female replacement stock) on perhaps 200,000 acres of moor. With fewer owners, shepherding has become more difficult. Those who no longer keep moor sheep often fail to keep up their fences against the moor (though a recent court case has confirmed their legal obligation to do so). Sheep which get into the intensive grass reserved for cows are sometimes dogged out very roughly, and an in-lamb ewe can suffer badly from being chased by a dog even if it is not mauled. An even greater problem is the vast increase of traffic on the open moor roads, where each year the slaughter of sheep and lambs is appalling. The main roads alone would require forty or fifty miles of fencing—destroying much of their beauty. The sheep themselves add to the scenic attractions of the open moor and there is a strong case for keeping the roads open and paying compensation for any sheep killed on them from some amenity fund. Motorists also need educating to the fact that open moor roads are not race-tracks.

The actual area of moor is considerably less than in Marshall's day. A good many acres, particularly towards the Vale of Pickering, were enclosed in his time. A good many farms on the moor above Hawnby and at many dale heads, where people scratched a bare living till the 1930s, have gone back to heather. Both the Forestry

Commission and private afforestation schemes have taken over a lot of moor in the last half century. The Commission moved into North East Yorkshire in 1921 to plant the first new forests above Pickering. Since then it has planted almost 40,000 acres in the Allerston and Hambleton areas. With private planting this has meant a loss of perhaps 7 per cent of previous sheep-grazing area. Many valley bottoms, such as Wheeldale and Rutmoor Gills, previously bracken-infested, have been planted, and where thick bracken land is taken this is almost certainly an improvement both economically and aesthetically, as long as forestry blocks do not cut off the moor tops entirely from the lower lands. The North Riding forests provide employment for 500 men with an annual wages bill of well over £300,000. This is more employment and more wages than sheep-grazing could have afforded. The annual increase in the value of timber is said to be now substantially greater than this wages bill. About 40 forest workers' holdings have been created with eight to ten acres of land each as part-time occupation. The Forestry Commission is doing a good deal to open up its forests for recreational purposes, establishing camping and caravan sites and forest drives or nature trails. It would be a great pity if there were any further forest encroachment on the higher moors but on the whole to date we must welcome the contribution forestry has made to the economy of the North York Moors.

In recent years there has been some controversy over the reclamation of large areas of heather moor—mainly in the Pickering–Scarborough–Whitby triangle—for agricultural purposes. No farmer can be blamed for making the best agricultural use of any land available. Traces of an early large-scale attempt at this by Sir Charles Turner on Kempswidden Moor near Kildale can still be seen. This —about 1790—was not very successful but at that time Marshall was suggesting a sounder system of doing it. There is no doubt that a lot of heather moor could be converted to good grass—at a cost which is only made economic by public subsidy. The national interest however must surely be to keep these wild tracts of heather in their natural state for recreational purposes. With rapidly growing population, and a very small amount of wild country available, it must be wrong to use public money against the public interest by subsidising reclamation of this sort.

In the dales below the moors there have been many changes in the last 25 years. Right up till the last war each dale was practically a closed shop where the same families had farmed for five hundred years. In Bilsdale, Garbutts, Ainsleys, Johnsons, Wards and Lengs can be found in 14th and 15th century records. Many of the farms—Ellermire, William Beck, Wether Cote, Gimmer Cote and Ewe Cote are mentioned in 1301. Since the war—and since the Duncombe Park Estate sold off many of its farms—there has been a great influx of "foreigners," and many of the old families have left or retired. In modern conditions the difficulty with Dales farms has been their comparative smallness. Many were of 50 to 60 acres, and 100 acres was quite large—and that would include some rough pasture in "intakes" at the moor edge. There is little room for expansion and improvement, and buildings are often totally inadequate for modern intensive farming. When the economic unit has become 80 cows or 100 breeding sows, amalgamation or co-operation has become essential. Mr. Jack Brunton who has recently bought the Urra Estate is planning—against some criticism—to run some of the farms together. To anyone who knows the history of these farms and the families that have farmed them over the centuries, this is tragic. But it is inevitable. Another successful invasion of Bilsdale has been made by Tom and Tony Archibald, brothers who farm 250 acres at Thorn Flatt, on the Cleveland Plain at East Harlsey. They first rented the 100 acre Laverock Hall in Bilsdale for rearing young stock, and were so impressed by the good grass in this sheltered, moist valley that they bought Thorn Hill and Oak House, each 100 acres, south of Chop Gate. In an interesting modern example of transhumance they move their 100 strong dairy herd up into Bilsdale for the summer grass, leaving the Harlsey grass for silage for the cows to come down to again in autumn. They have also built a 100 sow farrowing unit at Oak House, a large unit for any of these dales. Unless the dales are to be kept as museum pieces, with farmers paid a social subsidy as exhibitors of a wonderful but no longer economic way of life, this sort of development is bound to continue.

It still remains true that a farm with a good sheep stray can be an economic family unit. Four miles south of Guisborough is Sleddale Farm, a green island of 83 acres of cultivated land in a thousand acre sea of heather. It is a museum piece in one sense—"Sleddale Cote" was mentioned in 1300, an Iron Age village site is close by, and traces of Celtic cultivation underlie the modern fields. Cultivation has been continuous here for at least 2,000 years. Several people still remember Jimmy Davison who farmed Sleddale from before 1890 to 1912. Ralph Calvert, who died in 1966 at the age of 76 or so, worked for many years at Sleddale as a boy—it was his first place. He remembered particularly Davidson's dog, Charlie, as the best he had ever known: "It'd work a mile out o' sight." When Davidson fell ill he would work the dog from his bedroom window over a thousand acres of moor. At the sale in 1912 Ralph Calvert recalled that some good Swaledale ewes had made 23s. 6d., and some cows over £20. He and another lad could have taken the place over for £500—but they hadn't even enough money to buy the dog. "We tried hard, but a chap fra Stowsla cum wi a collar an chain riddy an seean ootbid us poor lads. Nine pund Charlie fetched—a lot o money for a dog i them days. But he was a grand dog—ah've nivver seen a better."

Joe Muir succeeded Jimmy Davidson, and to get started in farming he had to work during the day in the Belmont ironstone mine below Guisborough Highcliffe, working his land in the evenings after a three mile walk back. Eventually he ran 600 sheep on the moor and his son Robert, who now farms at Haggit Hill, Rounton, on the Cleveland plain, remembers with some nostalgia the years of his childhood at Sleddale, and the three mile walk to school, over the moor and through the woods to Hutton. In 1924, Frederick Petch Proud came from Bransdale to succeed Muirs at Sleddale. His brother Tom was keeper to the Earl of Feversham on the Bilsdale moors. Fred was a big man, badly wounded in World War I. As his sons grew big enough to be useful, a lot of work went into reclaiming more land from the moor. The contrast now is startling. At one side of the boundary wall is rough heather and bog, at the other green pastures and cornland, well farmed and well fertilized. The land was worked with horses—and breeding them was a profitable side-line—till 1948. Some young horses were sold to buy cows and the last horse died in 1968. Tractors did the major reclamation work. The Ministry allowed £4 an acre for ploughing out, although one intake of seven acres actually cost £65. Some of the reclamation cost £60 an acre—and it was years before good crops could be grown. There are now 430 ewes, 120 gimmer hoggs for future replacements, and over 300 lambs. On the 83 cultivated acres are 91 beef beasts as well as numerous geese and turkeys.

For centuries Sleddale's only approach for wheeled traffic was a rough cart track for three miles to Hutton—often impassable. In 1954, Fred Proud put a new slag road in direct across the moor to the Percy Cross road (for Kildale). Over 1,300 yards long, this took 200 loads of slag and cost £2,000. Some places swallowed several loads, and often a load filled only the place where it was tipped. Fred Proud died in 1968. The farm is being carried on very efficiently by his eldest son George (and his wife, Pat, a Durham miner's daughter) with the help of George's brothers Tom and Fred (who play cricket for Kildale and have done the Lyke Wake Walk several times!). Another brother, John, is a University agricultural lecturer. This is a little of one farm's history, where men have fought the moor for two thousand years, winning some, losing a little, winning it back again. The long fight against bog and heather must often have been a labour of love that went beyond economics—not love, perhaps, but sheer determination to win against the odds.

Another dales family provides an interesting study of farming development over two generations. The Wass family are mentioned in Bilsdale in 1485. In the 1880s, John Wass appears as tenant of Hon End in Farndale, an attractive farm-house well-placed on the western slopes of the dale just below Hon Nab, site of a Bronze Age camp. Seven children were born there in eight years. One, Willie, went to London as a draper and left £100 to Low Mill Chapel.

George went to farm at Wombleton Grange in the Vale of Pickering —his son, Harry Wass, farms that now and another son, John Potter Wass, came back into Farndale to Lowna. Joseph and Harry also farmed but had no family. James Todd Wass carried on at Hon End and had five children, all of whom are farming or have married farmers. I used to visit Hon End frequently between 1936 and 1939, after James Todd Wass had died. His three sons John, Harry and Fred, and two daughters Mary and Tamar, ran the farm as a family partnership. They had one lad living in, and used to kill seven pigs a year—one each, and one for visitors. This was still dales life at its traditional best. There was not much cash income in the 1930s, but there was plenty of happiness and comradeship, and the food was magnificent. There was a great hearth in the kitchen with turf plate and reckon-hook. Turves were cut from the moor above and sledded down to stack for winter fuel. There were always plenty of rabbits, hens, ducks, milk, eggs and butter.

Tamar married first in December 1938, and within four months all five were married. John, the eldest, stayed at Hon End for ten years, then retired. His daughter, Margaret, married a farmer, Brian Leckonby (another very old dales name)—and they have Ankness, an isolated moorland farm between Farndale and Bransdale, which was for many years occupied by another ancient family, the Wainds. Harry Wass now owns The Green, Fadmoor, 150 acres or more, with one son following on, one a mechanic and one a Methodist minister. Fred is in Farndale at High Wold House, producing milk on a hundred acres without any moor stray—one of the farms with a difficult future. Tamar married a farmer, John Atkinson of Lastingham (70 acres and milk). Mary married William H. Dunn, whose father was then farming Abbott Hagg Farm just above Rievaulx Abbey. Willie and Mary Dunn went to farm at another of the old Abbey granges near-by, New Leys, on the east side of the Helmsley–Bilsdale road. Of their family, Nicholas and Paul are helping to run the home farm, while the eldest son, Christopher, has married and is farming in Bransdale, at Breck House. A daughter, Judith, is at university. In this way one family has spread over a good deal of dales farm-land.

New Leys itself may be taken as a sample of a type of land which is not moorland at all, though generally classed with the area—the limestone plateaux which are higher than the moors below them at this point, but which are more akin to the Wolds in their farming system. The land is light, and there is not a drain on the place at New Leys. Though the soil is based on the corallian limestone of the Tabular Hills, it needs lime from time to time—as much as two tons every three or four years—because lime leaches quickly out of the cultivated soil. Of the farm's 196 acres, 50 are in corn—all barley. Yields may be 40 cwt. in a good year but the average is 32. This land can suffer a good deal from drought. There are 40 Friesian cows milked in a 6-stall, 3-point parlour, and 150 ewes—40 are Dalesbred, 15 are Pedigree Teeswater and 95 are Mashams, the Masham being the offspring of a Teeswater ram and a Dalesbred ewe. There are also eight very friendly goats which are used for fostering any lambs which lose their mothers. To give a little more detail about this farm, here are some extracts from Paul Dunn's diary:

30 June, 1969—Dad rotovating all day in fallow. I was turning hay. Grand hay day, but came on a shower at night. Nicholas went to Burnsides for a part for the Major (tractor). I cut three times round Hogg Field then broke down . . . 1st July—sew fertiliser on fallow (for soft turnips) and topdressed Seven and Four Acre with Nitrogen . . . 7 Oct. 1970—Dobson's lime men came —grand day but came a damp afternoon. Grinding and Mixing. Nicholas gathering wicks off Eight Acre . . . 1st April 1970—Terrible day. About 4 inches of snow, blowing into 4 feet drifts. Our road blocked. Milk recorder got stuck. Terrible for lambs. Got two sets of four—all lived. 150 ewes went to tup and 226 lambs have been weaned.

Ditching

5.   *The Vale of Pickering,*
The Wolds and Holderness

THE southern part of Bilsdale, Ryedale and the limestone valleys
and plateaux round Rievaulx centre on Helmsley as their market,
guarded symbolically still by the 12th century keep of the de Ros
castle, towering above its grass-covered mounds and moats. The
dales to the east—Bransdale, Farndale and Rosedale—centre on
the market towns of Kirkby Moorside and Pickering, each also once
guarded by ancient castles. Pickering has been an important regional
centre since the Middle Ages, when the Pickering Forest Court held
sway over a wide area. Between these towns on the lower slopes of
the Tabular Limestone Hills, and the chalk Wolds which rise to the
south, lies the old glacial lake of Pickering, 24 miles long and from
four to six miles wide—120 square miles of Ice Age alluvium on top
of Kimmeridge Clay. At the eastern end is some peat—and some of
our oldest habitation sites, at Starr Carr and Flixton, going back
perhaps 10,000 years. There are many sites of Roman villas.

On parts of the old lake edge, and on clay deposits, grass and
dairying may predominate—as instanced by George Scaling, at
Cliff House, Sinnington, who has a herd of 60 Friesians with 40–50
followers, on 90 acres of grass, with 130–150 acres of cereals.
Generally however this is an important arable area, and the Rilling-
ton district is famous for the fine quality of its barley. Previously
a good deal of fruit was grown, but though the orchards are more
luxuriant here than elsewhere in Yorkshire, there is little grown on a
commercial scale. Here, as in Yorkshire as a whole, the acreage
under small fruit has declined considerably in the past decade, though
as many as 14 tons of gooseberries have been sold from the Vale for
jam-making in Leeds. There is a large nursery famous for its roses—
Rogers of Pickering—and one or two smaller ones. There is also an
important agricultural engineering works, with a large export trade
—Russells of Kirkby Moorside. At Kirkby Misperton is a rare form
of agricultural exploitation in the shape of an increasingly popular
zoo! East of here, the carrs and ings and marishes of Allerston,
Yedingham, Snainton, and Seamer—probably the last to be reclaimed—form a peculiar low flat landscape between the limestone
hills to the north and the wolds to the south. In recent wet years the

water table has risen so much in parts of this Vale that some fields have had to be taken out of cultivation and put back down to grass, but much is good wheat land.

From the Vale the Wolds rise abruptly to over 500 feet at Heslerton Brow and Staxton Brow. Some of the Wolds are over 800 feet and they can be very bleak. This is a region of large farms and large fields, with the farm-houses, few and far between, always sheltered from three sides by belts of trees. In earlier times this was poor land, with sheep the main feature. In the early 19th century rabbits were farmed in quite a big way, in warrens of up to 2,000 acres, and crops of grass and turnips were sown for the rabbits. Only the fattest were sold, in Hull and York and the industrial towns of the West Riding. Stamford Bridge and Malton were local markets for the skins. The Sykes family of Sledmere did more than most to improve Wolds agriculture — particularly Sir Tatton 1771–1863, who liked to act as his own drover (and dress for the part) when moving sheep. Improved rotations, with break crops—particularly turnips, on which the sheep were traditionally folded—led to the Wolds becoming mainly arable, and famous for their barley, though oats have also long been a popular crop. The Wolds became one of the main regions to adopt Bakewell's improved breed of Leicester sheep and have been the last stronghold of the breed though Down crosses have largely replaced them even here. Stock on the Wolds used to be limited by the scarcity of water—often dew-ponds were the only source for animals. Piped supplies have widened the scope of Wolds farming so that there are even some dairy herds as well as intensive pig units and the like.

At the top of Garrowby Hill, 800 feet, Jack Megginson farms Cot Nab, 480 acres. Almost half this is in barley (220 acres) and about 100 acres is taken up by narrow, steep-sided dales which are little more than rough grazing. There is a flock of 140 Suffolk cross ewes, not usually folded in the traditional way, though an uncle not far away has a flock of 500–600 mainly folded on turnips in the winter. There are also 30 acres of seed potatoes. Below Cot Nab— 700 feet below—on the heavy wet clay at the foot of the Wolds on the west, a close friend of Jack's, Bob Sleightholme has 220 acres at Church Farm, Youlthorpe, of which 120 acres are corn and 100 acres intensive grass for an eighty cow dairy herd. This is a very modern unit with the latest kind of low-level herringbone parlour and automated feed rationing. There is also a pig unit for 100 breeding sows. Stock, and assistance, are sometimes exchanged between these two very different farms only a mile apart.

To the east the Wolds merge into the cliffs between Filey and Bridlington, where the chalk goes under the clay and coastal drift. This is the country of the "gypsies" (hard g), those mysterious streams which appear only at certain seasons, sometimes after several years. The most famous is *the* Gypsy Race which runs a dozen miles or so down the Great Wold Valley to Bridlington. According to old records, there have been periods when this stream has only run five times in twenty years, and there even used to be some superstition that when its "woe-waters" ran, there would be trouble. On one such occasion, in 1795, a thunderbolt fell on Wold Newton! There is a simple geological explanation of these springs and streams. Clay underlies the chalk. The chalk will hold a lot of water and it is only when it becomes saturated that water will come to the surface at all. It is a significant comment on recent weather that summer 1970 was the first time the Gypsy Race had been dry for four years, though before that it had been dry every summer for a long time.

John Stephenson farms land on both sides of the Gypsy Race—250 acres at Wandale Farm on the old Roman road over the Wolds, the Wold Gate or "Waud Yat"—and 230 acres at Eastfield, near the old village of Easton which the Brontë sisters used to visit. The Stephenson family have been at Wandale for 80 years—three generations, but John had five years in the merchant navy. He is a P & O navigator with three round the world voyages to his credit and he certainly likes everything shipshape and Bridlington fashion. The land varies from heavy clay to light chalk—and the heavier land tends to be at the top of the slopes. John used to grow 80 acres of potatoes, but his maximum yield was 12 tons. His brother, David, at Market Weighton could yield 20 tons, so after a few years on barley beef John Stephenson's main enterprise apart from cereals is milk—from a herd of 100 Friesians, in cubicles, with self-feed silage from covered clamps. The cows, with 22 replacement heifers, occupy about 108 acres of grass. There are 350 acres of cereals, and this farm exemplifies the point that for really efficient cereal growing something like 300 acres is now the minimum size. With interest alone at over 10 per cent, no less an enterprise can afford a new combine costing over £3,000. There is a continuous drier which can handle 45 acres or 90 tons of corn a day—the average cut in the 1970 harvest was 30 acres a day. Harvest was over in twelve days. There are 100 acres of winter wheat, the rest barley and to make his system pay John Stephenson relies on getting two tons—at the very least 38 cwts.—average yield per acre. All the land is chisel-ploughed twice, disced, and cultivated with spring tines right up to the hedge bottoms before being finally ploughed for the winter.There is not a wicken to be seen. Only two men are employed on the arable—with good wages and overtime.

As you go south from Bridlington there is some very good farming to be seen. Trees gradually grow fewer and smaller till you are in the bare landscape of Holderness. The North Sea is always close—and getting closer at the rate of five feet a year as the cliffs get washed away. The soil is mainly medium to heavy boulder clay, with bits of morainic sand and gravel and old glacial lake bottoms of alluvial land. In the Middle Ages most of the area was swamp. Much drain-

ing had been done in Arthur Young's day and in 1830 Cobbett
wrote one of his purple passages about the area—"I used to wonder
that Yorkshire should send us of the south those beautiful cattle with
short horns and straight deep bodies. You have only to see the
country—on the north side of the mouth of the Humber—to cease
to wonder at that. It is as flat and fat as the land between Holbeach
and Boston." And, with the exception of the Holbeach lands, he
had seen no other in the finest parts of England to compare with the
land on the banks of the Humber, the Holderness country included.

The best of Holderness may have been grazing pastures in his day,
but soon after it became largely arable, as it is now. Perhaps the most
fertile part of all is Sunk Island, almost entirely warp formed by the
tides which have swept soil down for generations from higher up
the coast. Until the end of the 18th century it was still an island but
later deposits have entirely filled the channel. In 1921 one farmer
weighed up 450 cwt. wheat from a 10-acre field when 20–25 cwt. an
acre was generally considered good. Good yields of potatoes are
obtained from the warp, and mustard has also been grown for
Colmans. Like all heavy land districts Holderness suffered badly
in the 1930s, and 400 acre farms were sold for less than £3,000.
They would fetch at least £120,000 now.

East of the sand and gravel pits of Keyingham, on a ridge of
medium heavy land beyond the RAF radar station of Patrington
where there was once an Elizabethan warning beacon, lie the
thousand acres of land farmed by the Hird Brothers of Holmpton—
Harry, George and John (though the last two have now retired).
The family has been at West Farm, Holmpton, since 1913. Harry's
father, after shepherding for 15s. a week, took to marketing and
dealing, then to auctioneering, and was for fifty years at the Hull
Cattle market. He bought West Farm in 1918, and two other farms
were added later—there are now 750 acres arable and 330 grass.
Here is still grown another traditional Holderness crop—beans
(50 acres). They also still grow 50 acres of oats, and there are 360
acres of barley—the rest is wheat. There are three herds of Friesian
cows, in three different parlours each with its own bulk tank—
altogether 190 cows and 30 followers. With 140 cows in milk, 13,000
gallons of milk were sold in August 1970 and the milk cheque was
£2,000. There are 200 Scotch Half Bred ewes (Cheviot X Border
Leicester) run with Suffolk tups and there are 50 sows—weaners
being sold at 10 weeks. Harry's father was a strong believer in
"yowes, sows an cows!"

Apart from son and grandson, there are three cowmen and four
or five others—staff has been cut down in recent years and overtime increased to give higher wages. At one time this land was
worked by 40 horses—now there are eight or nine tractors. From the
ridge top you can see the North Sea much too close on one side—
Harry has seen a whole field disappear during his lifetime—and on

Contrasts in cattle. Top: Milking time for a prize Jersey herd at a farm at Skirlaugh, near Hull. (John Edenbrow). Bottom: Friesians grazing on highly fertile grassland near Patrington. (W. R. Mitchell).

Grange Farm, Girrick (see Cleveland chapter). Above: Mr. Joe Sumley with a Cleveland Bay. (Wm. E. Grier). Opposite, top: Mr. Sumley and the author survey the Jersey herd at the farm. Opposite, bottom: A general view of the farm and its approaches. (both Jim Larkin Fotos).

Bringing in bracken for bedding in Garsdale. (Bertram Unné).

Hauling and Walling

A mixed load of sheep, milk and sticking being taken over the moorland road between Wensleydale and Swaledale. (Ron and Lucie Hinson).

Fillings being inserted during dry-stone walling operations in Bishopdale. (John S. Murray).

A Swaledale farmer repairs the ravages of winter gales. (Ron and Lucie Hinson).

Scenes from the past. Above: Pause for refreshment during ploughing
near Thixendale, East Yorkshire. (Bertram Unné). Opposite, top:
Harvesting oats near Goole on a September evening. (J. A. Winterburn).
Opposite, bottom: An old Cleveland horse rolley used as a temporary
milk stand on the Guisborough-Whitby road during a foot-and-mouth
epidemic. (Jim Larkin Fotos).

A classic view of a moorland farm at Grains o' the Beck, Upper Teesdale, vividly illustrating hard-won reclamation. (Geoffrey N. Wright).

the other you can see across the dip of the Humber to the lights of Lincolnshire. Generally in the area, says Harry, dairying has decreased and arable farming has increased—sometimes in the immediate landscape to the north of here you can in winter see nothing but ploughed land, without an acre of grass—and there have been a lot of big pig and poultry units started.

6. *The Central Plain of York and the Deep South*

FROM Blackwell on the Tees to Bawtry on the old A1 is just over 80 miles. Eight miles west of this line Yorkshire dips eight miles further south towards the Nottingham sandlands, while the Tees bends further north. Ninety miles is a vast stretch of country to cover as one geographical or agricultural unit, yet, after only a mile or two from the Tees, all this drains into the Humber and there is a certain unity about it, though there is a great diversity of soils and of farms.

Throughout the area however there is one type of farm which is frequent, the "cash roots" farm. These generally have 25 per cent of their area in cash root crops—potatoes and sugar beet mainly, with carrots occasionally (much less in recent years). Sometimes there is diversification into market garden crops like cabbages and sprouts, or even in the southern part of the area into peas for canning. The rest of the area of these farms is corn, with just a small proportion under temporary grass, and there may be sheep or bullocks to eat this off, though some have no stock at all and the hay is sold.

Although the "Vale of York cash roots farm" is a well-known type, there are mixed farms throughout the area, and always the specialist dairy farm not far away. Towards the north of the area, on light land just south of Ripon, Cyril Wade has 120 acres all grass near Littlethorpe. He bought his first four heifers in 1949, when he had little more than 10 acres, and has built up to 60 cows and followers as a part-time farmer, being for most of the time connected with the engineering side of the family firm of worsted piece dyers. He would however claim to be a farmer who had been reluctantly tied to industry for too long, rather than an industrialist turned farmer, and now in his later sixties is happily engaged in full-time farming, turning out about 50,000 gallons of milk a year. Mrs. Gwen Wade is an artist and writer, and member of Council of the Yorkshire Dialect Society. She has edited two anthologies of dialect verse for the society and one of her own verses "Unharrowed Ground" might be quoted in part here to illustrate the problem of weeds on light land!

> *Ther's a bank in a pastur*
> *Close bi ahr spot*
> *At natters t'young farmer*
> *A hell of a lot;*
> *For it's wick wi field-pansy,*
> *Wi pig-nut an tansy,*
> *Burnet an buttercup*
> *Bonny as owt!*
>
> *Ower brant for his sprayin*
> *An poor for his plew*
> *Is his bank wi its clutter*
> *O' blooms keekin through.*
> *All yon muckment!-Field-pansy,*
> *Wi pig-nut an tansy,*
> *Burnet an buttercup,*
> *Bonny as owt!*

South of Ripon, the valleys of Yore and Nidd are just merging with the Plain of York. There are large arable farms, but there are also small dales-type mixed farms. One such is Cayton Grange, near Ripley, where I used to work during school holidays in the early 1930s. It was a 60 acre, two-horse spot in those days, lost in a wooded valley five fields away from the main Ripon road. John Fryer (who died in 1936 at 73) had been born at Ringbeck, Kirkby Malzeard, and served his time as a joiner but he couldn't stand pitchpine. At 22 he took over Prospect Farm at Hartwith in Nidderdale—where in 1922 as a boy of 7 I first met the family. He had married Grace Marshall, whose father, Will, was a working shareholder in a lead mine at the head of Nidderdale but his team struck an artesian well and went bankrupt. John Fryer was an expert judge of horses and always had a couple of good ones. There were six or seven cows, butter and eggs were sold, and a few beasts reared. I used to turn away at the handle of the cream separator in the dairy—we were too far from civilisation to sell milk in those days. Outside there was hay-making and turnip hoeing. Mrs. Fryer was crippled by rheumatism and scarcely ever stirred from her chair. But from that chair, strategically placed between the kitchen range and the table, she presided over everyone. She was a wonderful cook, and Cayton Grange was certainly what farm lads used to call "a good meat house." Dinner always began with a savoury pudding —suet or Yorkshire. There was always a pig or two being fattened, sides of bacon and hams hanging in the kitchen, and the valleys swarmed with rabbits—snaring them contributed not a little to the farm's cash income, which must have been pitifully small. The old couple died and the sons sold out before times got better—and I doubt if the whole sale of stock and machinery brought as much as one would pay for a second-hand combine now.

I visited Cayton Grange recently, after nearly forty years, not knowing what I should find. It was exactly the same—the old cart shed, the cobbled yard, the stables still intact, some chains on a nail that might have been there all the forty years. I hadn't thought there was anywhere left so unaffected by the improvements or the ravages of time—untouched alike by progress and by decay. The tenant, a Mr. Kirby, had come here after the Fryers left 33 years ago, with the land reduced to 30 acres. He kept half a dozen suckler cows and their off-spring, and a son and daughter worked elsewhere. This small farm had been used as a pleasant home, paying its own rent perhaps but not expected to provide the main income, and in this way it had survived as other small farms may and should survive.

One of my oldest friends is Edward Robinson, with whom for a year at the age of seven I went to the village school at Hartwith. Edward's father, Mark, was one of those countrymen who can turn their hands to anything. He would take contracts for ditching, hedging, walling—he must have put up miles of dry stone walls—he repaired houses, laid concrete when old cowbyres had to be improved, and dug graves. He needed every penny because all he had was a 36 acre holding on the moor edge near Brimham Rocks, with two or three cows for suckling calves. With carefully saved capital he moved the family down to the central plain of York in 1936—to a County Council smallholding of 50 acres, Woodside Farm, Acaster Malbis. In 1939 Edward took over a new Council smallholding nearby—the 50 acre Foss Field. The West Riding County Council has provided a number of such smallholdings—many in the York area and between York and Boroughbridge—under an Act of 1908, as have the North and East Ridings. Many were allocated to ex-Servicemen in the 1920s, but others provided an opportunity for outstanding farm-workers to become farmers on their own account. They are usually 50–60 acres, though most tenants have bought or rented a little extra land in addition. The Central Plain farms, well managed, have been reckoned to produce an income of about £2,000 a year to cover the farmer's labour, management and interest on capital—and the working capital required now, even with second-hand machinery, is probably at least £5,000, though a man might start with half that sum. There is still a big demand for such holdings, the rent of which is usually £6 or £7 an acre, but they are frequently criticised in present conditions for being too small—particularly for being too large for one man to work satisfactorily, but not large enough to employ a full time assistant at present wages.

Edward Robinson's farm is a small edition of a cash roots farm, with a small dairy enterprise. The ten Friesian milking cows take up the grass break and have in the past paid the wages of a full-time man. The main enterprise is corn—10 acres wheat, 10 oats, 30 barley

—and there are seven acres of sugar beet. Edward has bought another 21 acres giving him 71 altogether. Part of his land is one of the ancient strips of meadow in an Ings field by the tidal Ouse. Several farmers have strips here; there are no boundaries in the field, but stones at each end (except that most of those at one end have dropped into the river!) A rate is levied among them for fencing, and stock may be turned in at the rate of one beast to the acre after the hay is taken off. Edward himself was Ings master last year.

Foss Field may have been cultivated by the Romans. The land is medium, with clay and sand in layers and no great depth of soil, though over towards Cawood the soil is more warp-like. A mile north on the Ainsty ridge of Askham Bryan (where the West Riding Agricultural College is situated), it is more gravelly and stony. With the latest increase in agricultural wages, stamp and pension, Edward sees little prospect of being able to employ a man full time as in the past—the man would be getting more out of it than he would. The alternative is to cut down production a bit, take things easier, and use casual labour when necessary. There is a certain amount of co-operation in labour and machinery use among neighbours. Of Edward's brothers, Charles continues at Woodside Farm (their father died in 1960) and has acquired a hundred acres of his own in addition, while Arthur, helped by his father, was able to buy the 160 acre College Farm at Acaster Selby and is well known for his British Friesians. This is not bad from the small and frugal beginnings of the moor edge holding at Hartwith. By one of those farming coincidences, Edward's wife Eleanor, who was also a Robinson, is related to the Prouds who farm at Sleddale.

At the other side of the Ouse is the Escrick moraine and the soils —sometimes clay, sometimes blow-away sand—that slope to the Derwent. Mr. Sydney N. Patrick came to Lawns Farm, North Duffield, in 1938—and sold Arran Banner potatoes for 45s. a ton. The 230 acres were mainly arable then, but after the war he and his sons went in more for grass, with a milking herd of 40 Friesians. In 1958 these were on a very good self-feed silage lay-out, but later the Patricks ran into trouble with effluent draining into the Derwent, and returned to hay feeding. The two sons, Stanley and Leslie, now each have a 70 cow herd and just over 200 acres each—the one at Lawns Farm and the other at Springfield House. In addition they rent Horn Farm, Thorganby, 130 acres, in partnership. In spite of these dairy herds, the prevailing type of farm in the area is still cash roots, with little grass. Towards Selby there is a good deal of market gardening on silts and sands.

Where the Ouse turns east towards the Humber, the Drax Power Station chimney rises to 850 feet above these flat lands, dwarfing the beautiful spire of Hemingbrough Church opposite. Wood House Farm at Drax was sold for £6,750 (186 acres) at auction in 1941 when it was advertised as "an accredited dairy farm for 80 cows, one of the finest and most up-to-date in this part of Yorkshire." There is no Wood House Farm now—it is all Power Station. Mr. George William Dunn of Drax has worked on farms in the area all his life, and he is now 65. Some warping was still going on in the area when he was a young man, and there is a lot of good warp by the Derwent, though there is also some badly drained low land which is worth little. He works now at Rusholme Grange, a 400 acre farm along the south bank of the Ouse where much of the land is 3 feet warp on top of clay and the drains are over 5 feet deep. This is rich land that was warped a century ago. Some of it will grow wheat for three years at two tons to the acre, and some oats in 1970 yielded 52 cwts. to the acre. Usually they grow two years wheat and a three years ley—and the leys I saw in October 1970 were the finest and heaviest late grass I have ever seen. Silage is made, and 60 beasts fattened on it.

South and west of Goole is the Dutch River country, drained by Sir Cornelius Vermuyden in the 17th century when he cut a completely new channel for the river Don. For most people this is a desolate and un-Yorkshire landscape. On large areas of land farming seems to be a very impersonal business proposition—but no-one would live here for the pleasure of it. Just across the border is the Isle of Axholme and Epworth (where a bit of the old common field and strip system still survives). Between Snaith and Thorne (once famous for their "Marshes" and "Wastes") and the warp villages of Swinefleet, Reedness and Adlingfleet, is a vast area of arable with few habitations. The Brabbs family farm at Hatfield and Tudworth Green—500 acres or so, again mainly a cash root system of sugar beet, potatoes, barley and wheat with some beef fattening. The land is medium sand with some peat—peat is dug for horticultural use nearby—but there is also clay. Some clay in the area is so strong that unless it is ploughed by November and gets a frosty winter it is unworkable. Some of the sand turns to a grinding paste and wears out ploughshares in an afternoon. Towards Bawtry are sand and gravel diggings. One man here remarked, "I'd rather hev it clagging ti mi boots than blinndin me i' t'ee."

Eight miles to the west you are back in something more like Yorkshire again. From the Maltby area came Fred Kitching, a farmworker all his life, who in *Brother to the Ox* gave a vivid picture of life in this corner of Yorkshire. The long ridge of magnesian limestone which forms the western edge of the central plain is quarried extensively north of Maltby. At its southern end the land tends not to be so good as further north—it is thinner, and more liable to drought. At Braithwell was one of the only two raddle mines in England. Raddle or rud is a red ochre stone used, either as a stick or a powder, for marking sheep. It was also used as an early form of distemper for walls, and by the courtesan for raddling her cheeks. The raddle, a sort of haematite, is found in the coal measure immediately below the great yellow limestone. The Braithwell/ Micklebring mines and mill closed during the Great War, though some local farmers still use their own raddle. One man who is an authority on the old industry is Mr. Albert Dunstan who with his son Joseph and grandson Peter farms 72 acres at Braithwell, though only one acre is attached to the farmhouse and buildings in Braithwell village. The Dunstans have been in Braithwell for seven generations, and some of them got land when the commons were enclosed in 1838. Orchard House itself goes back to 1500, and the byre to 1688. Despite the drawbacks of living in the village with no land attached, the Dunstans manage to keep 52 cows, and young stock, on the 72 acres. The cows have to be kept in at nights even in summer because of the long and awkward journey to the fields. For the same reason hay has to be made, not silage. About 20 acres of barley is grown and stored in a wet grain silo, and more is bought. The limestone here tends to drought off in dry weather, and the clay parts of the farm produce better grass. There is very rarely any magnesium deficiency in the cows, but after a bout of milk fever a few years ago they did put lime on some fields. Average milk yield is round 900 gallons.

Further north again, the magnesian limestone is richer. Messrs. G. Pick & Sons of Old Hall Farm, Scarthingwell, Saxton, have 780 acres bordering on the old battlefield of Towton. This includes about 100 acres of ings on the Wharfe near Ulleskelf which is good grazing for fattening bullocks. Otherwise it is a typical limestone farm with 40 acres potatoes, 35 acres sugar beet, 300 acres wheat and 240 acres barley. They no longer keep sheep or grow roots for folding, though some limestone farmers still do this, mainly with Down or Down Cross sheep. Just next door however is limestone of a poorer character, badly drained, but growing good grass, and here Mr. G. Chapman has a herd of dairy cows. Average yields on the limestone here are 32 cwt. for barley, 12 tons for potatoes and 16 tons for sugar beet.

7.

The Industrial
West Riding

UNDER the shadow of the Parkgate Steel Works on the north side of Rotherham, Mr. C. Pearson and his two sons farm over 800 acres of coal measures round Greasbrough. Farms include Town End, Home Farm at Wentworth, and Peacock Lodge. In 1777, William Bray F.A.S. in his *Sketch of a Tour into Derbyshire and Yorkshire* wrote of the Marquis of Rockingham's agricultural improvements round Wentworth, "His draining of wet lands, his cultivation of turnips and introduction of the hoe, the new instruments he brought into use and the improvement of the old ones, will bring him the most lasting honour." Wentworth House had 365 windows, and at one time 40 grooms. Now it has 300 P.T. students and the Pearsons run four acres of the gardens and a shop commercially, chrysanthemums and tomatoes being important crops. On the farms there are 200 acres wheat, 400 acres barley, 35 acres potatoes and 200 acres of parklands. There are 10,000 head of poultry, 40 breeding sows, and 100 beef cattle, calves being bought in. The dairy herd was given up a year ago. There are also 120 Northcountry ewes run with Suffolk tups. Of this land, 300 acres has been surface mined for opencast coal and restored. It is bearing crops again, but not as before, and it takes a lot of fertiliser to keep it going. The problem here, as throughout this chapter, is industry with its pollution and the encroaching town.

North West again, beyond the Barnsley mines, Cawthorne seems much more rural than Greasbrough, but here also much land is only slowly recovering from opencast mining. As at Wentworth, one or two large estates have provided some bulwark against the tide of suburban development. The Charlesworths were tenants of Hill Top Farm, Cawthorne, for 300 years. The farm was just 50 acres of goodish land—typical of many in the green valleys outside the West Riding towns. From 1915 Douglas Charlesworth was mainly responsible for the farming, though his father was still alive. Like Orchard House at Braithwell, Hill Top had only a two-acre garth attached, and the land was scattered round the village. It can never have provided a very good living, but it is clear from Douglas Charlesworth's diary kept from 1915–1919 that it gave him a very good life. He enjoyed his farming, and he liked hunting and shooting —as well as swimming and skating. But his main interests in life were writing dialect poetry and studying local history. Here are a few extracts from this fascinating diary:—

March 30, 1915: Sowing oats, Tartar King, in Looby Field. Did this day buy the History of Worsborough at Mr. Hudson's in Shambles Street, Barnsley. Resolve that the growing habit of buying books must be checked having bought 30 in 3 months.

April 1st: To Barnsley with the milk as I do every morning—the children calling and trying to make me into an April Fool. People desiring much milk to make into hot cross buns. Talked with Mr. Percy Jackson of Skelmanthorpe on the feeding of cattle and production of milk, he arguing that it was better to buy hay at £3 per ton than to grow it. This evening sat reading the Complete Grazier and Norse Folk Tales.

April 7: Leading wood from Barkhouse Lane. In the evening pruning my roses then at dark reading Menzel's History of Germany by a great fire of oak logs.

April 9: At the French class at the museum. Lent W. Midgley my Lloyd George Land Report. Then to practise some bayonet exercises. Took another load of swedes to Redbrook.

October 16: On Dunford Moors—very foggy so that we only got one brace of birds on Grains Moss where I have seen us get 65. At lunch asked Metcalf about the Penistone breed of sheep—they were white faced horned sheep but are not so much kept now being superseded by Lonks.

November 13: Went pheasant shooting in Deffer Wood. A bright hard day with 4 inches snow on the tops. Birds plentiful and sport good.

The diary is a pleasant mixture of farming, of reading and talking, and of the war in the background. Douglas Charlesworth was a crack shot but was rejected as C3 for the Army. His local rifle team heavily defeated an army team in a shooting match and he was very worried about the bad shots who were due to leave for France shortly. He got £25 for a cow at Barnsley Auction, grumbled that commission had gone up from 2s. 6d. a cow to 3d. in the £, and bought a new Hornsby mowing machine for £13. He went to meetings of the Numismatics, Naturalists and Dialect Societies. He mourned the loss of Gypsy—"the best and truest dog that anyone ever had. She was unequalled as a cattle dog and worked very well with the gun." Douglas Charlesworth's poems show the same deep appreciation of the countryside and its past and of the country life around him. In one on ploughing (reproduced overleaf), he imagines all the generations that have ploughed the field before him, and their ghosts come to see how he gets on.

Douglas Charlesworth joined those ghosts himself in 1941 when he was only 57, and his son Martin had to take on all the work of the farm when he was 16. But long before then the kind of life which Douglas Charlesworth had enjoyed in his earlier years had ended. Martin worked hard at Hill Top for ten years, retailing milk and eggs and potatoes in the growing village and becoming attested. But by 1951 the land was all going for building or for opencast mining, and he was the last tenant. He bought the 61 acre New House Farm

at Birdsedge, and through the difficult '50s struggled up to 16 cows.
It was not sufficient, and by 1966 he had had enough of running this
single handed. He sold out, and took his first holiday in fifteen years
—a fortnight in Devon! Now he has a 15-acre smallholding of his
own, works full time as stockman on a nearby 250 acre farm, and is
much better off in every way. Other small farmers in the industrial
West Riding have followed this example, and probably many more
ought to.

Up, mi lad, it's time for ploughin,
　　Winds o mornin softly sweep,
Time at swingletrees were clinkin
　　Ovver t' furrows streight an deep.

But remember as tha farest,
　　Stridin t' stubble or green ley,
Ladt there are at ploughed afoor thee
　　I's'lang years at's goan for aye.

An they come, these shadow-ploughmen,
　　Sliven quietly i t'morn
To see hah tha frames at ploughin
　　T'fields they skeathed foor tha were born.

Silently they watch thee strivin,
　　Settin rig an drawin furr,
An they listen, still an wistful,
　　T'slipe an t'colter's steady purr.

Shap thisen, for tha art livin,
　　Step more gradely, whistle shrill;
T'skylark sings his song aboon thee,
　　T'sunbeams glitter upo t'hill.

When t'last furrow's felled an soiled,
　　An t'team stamps up t'claggy loin,
Breet een wait wi lang looks tender,
　　Smilin lips their welcome join.

T'failin land-fooak, still an quiet,
　　Shadowless on t'greyin hill,
Howd their heads up, smile contented
　　At there's maister-ploughmen still.

　　　　　　　　　　　　　—Douglas Charlesworth

Historically, what are now often the "problem farms" of the urban fringe began as cow-keepings, retailing milk in the towns. Often they had a "flying herd"—they did not do any rearing of young stock, but bought in newly calven cows, fed them well to get as much milk as possible, and sold them dry for beef. Though there are many snags with a flying herd there is nothing basically wrong about it. The main problem is size, coupled with increasing costs, and modern hygienic requirements for milk. These days, even if milk is retailed, to make a reasonable living from the one enterprise a minimum of 35 to 40 cows would need to be kept. On this sort of land this means 70 or 80 acres—and the necessary buildings. A great many of the farms were only 15–30 acres. Some of these have been amalgamated, some of the larger ones struggle on. The higher they are, the more difficult living becomes, and ten years ago on the moor ridge West of Elland towards Buckstones Moss I found a whole area derelict, the fields going back to moor, and only sheep wandering over the broken walls.

Nevertheless, in spite of the difficulties, and in spite of the acid soil and atmosphere, some farmers on a reasonable acreage using very intensive methods have had outstanding results. Some of these are I.C.I. costed farms. No survey of Yorkshire farming would be complete without a mention of the contribution made by several large firms towards its agriculture, either by research and advice, or by running their own experimental farms. The B.O.C.M. pig and poultry units, and their beef and milk progeny testing units at Barlby, did great service to farming in general and particularly to the county. Silcocks, and Barker Lee Smiths of Beverley, have also contributed a good deal. Such firms have frequently co-operated with the Agricultural Colleges—both at Askham Bryan and at Bishop Burton (East Riding). The I.C.I. have farms of their own on the Eston Hills in Cleveland but they have also provided detailed advice and cost accounts for several Yorkshire farms, in all areas. It is fair to say that these tend to be reasonably sized units with a fair economic chance, and that the farmers concerned were perhaps better than average to begin with if only in their readiness to act on advice. Continuing advice, and the benefits of a clear accounting system measuring the profitability of every department of the farm and perhaps every field has helped a great deal, even where success is mainly due to the farmer's own initiative and determination.

One farm that has been held up as a model for the industrial Pennines is that of Oliver Barraclough, Well Heads, Thornton, near Bradford. His 130 odd acres are at 1,000 feet on the edge of Denholme and Oxenhope Moors, with a rainfall of 45 inches. The soil is millstone grit, about a third of it being medium loam overlying gravel and the rest a heavy clay which poaches badly and has required a good deal of expensive draining. On this acreage a herd

of over 90 Friesian cows is kept, milk being sold wholesale—
£147.4 per cow in 1970–71, with a total expenditure on bought-in
concentrates and other purchased feed of just under £30. On the
other hand an average of 393 units of nitrogen, 45 units of phosphate
and 89 units of potash were applied to every acre—a total of nearly
90 tons of fertiliser. The whole farm is grass, with grazing controlled
by an electric fence moved twice a day. Grass for winter is all con-
served as silage in self-feed clamps—about 900 tons of it, in two cuts
(May–June and July–August). Thanks to heavy fertiliser usage, full
grazing is usually obtained even at this altitude until late October,
with day grazing into November. A few tons of hay are bought to
offer to cows during wet spells in the autumn before they come on to
silage. The average yield for 1970–71 was 803 gallons per cow and
total milk production 73,070 gallons.

This is intensive farming in difficult conditions. As a basis for
comparison, on the much better land of the Cleveland Plain I
produce a similar amount of milk on the same grass acreage from
slightly fewer cows for less than half that expenditure on fertiliser.
It would be safe to say that at least 50 tons of the fertiliser used—
or well over £1,000 worth without subsidy—is needed merely to
correct the geographical and climatic disadvantages of the area. Well
Heads is run by Mr. Barraclough and his son, with one man, and has
been costed by I.C.I. since 1947. No calves are reared but there is a
subsidiary poultry enterprise.

Another of the I.C.I. costed farms is Moor End, South Crossland,
Huddersfield—Mr. A. Sykes. This is a similar size—128 acres—in
perhaps slightly more favourable conditions, at 750 feet with 40
inches rainfall. The soil, though basically millstone grit, is light and
easily worked. Since this land does not poach easily, cows can be
turned out from April to November. Mr. Sykes won a Yorkshire
Grassland Society award for silage in 1967 and 1968 and a Grassland Management prize in 1969. He has made silage for 20 years.
In 1969–70 there were 94 cows (Ayrshire/Friesian) with an average
yield of 1,012 gallons. Fertiliser use has tended to be 15–20 per cent
less than that of Oliver Barraclough, again a measure of the rather
better conditions. Moor End is one of the many farms round
Huddersfield which still bottle their own milk, though bigger than
most. Many smaller ones could not survive without the higher price
for bottled milk, and some who cannot bottle have to work part-
time on other jobs. To get and keep labour, wages have to be com-
petitive with local industry, though this is true of much of Yorkshire. Like others in the area, Mr. Sykes used to grow cereals,
potatoes and swedes, and to keep a few sheep. In the last fifteen-
twenty years costings have shown the low profitability of these
enterprises compared with intensive milk on the right scale.

Three other outstanding farms are further south. The Thornley
Brothers farm Blacker Grange, Hoyland, near Barnsley—98 acres

of grass and 15 acres of rough land in the heart of the Yorkshire Coalfield. The heavy coal measures soil is difficult to manage, poaching badly when too wet and baking hard when too dry. A herd of 118 milking cows is kept on a paddock-grazing system— that is, two-acre paddocks are made, usually by putting permanent electric fences in larger fields but always giving the cows access to water. They are then given a whole paddock of fresh grass after each milking—so that, as at Blacker Grange, where there are 29 such paddocks the cows come back into the first one again after a fortnight and the grass has grown fresh for them. At Meadow Farm, Whitely Wood, Sheffield, Mr. D. Sanderson has 49 acres at 850 feet with another 23 acres further west into the Pennines. On this he keeps about 70 head of pedigree British Friesians, of which 30 are milking cows averaging 1,184 gallons. Again there is silage making and paddock grazing. This may be said to be about the minimum size for a satisfactory income, and then only because of such skilled and intensive management. At Grange Farm, Snydale, near Pontefract, Mr. J. Cressey has just over 100 acres, with the rainfall now dropped towards the central plain level—26 inches. He is leaving this farm shortly but has recently had a 68 cow dairy herd on just under 80 acres of grass, with 24 acres of barley and an intensive barley–beef unit of about 40. About 50 tons of fertiliser a year have been used, and 60,000 gallons of milk are sold.

These are examples of what is necessary for a successful farming enterprise. Anything much less can only yield a poor income in return for very hard labour—less than could be obtained working for someone else, though many Pennine farmers value their independence above all else. The industrial area should not be left without mentioning the liquorice which was once grown near Pontefract, though never on a large scale. The soil in a particular area of division between the red sandstone and the magnesian limestone—a deep, light loam with plenty of humus left by ancient floods and decaying matter—was particularly suitable for this deep-rooted crop, which was introduced by Black Friars in the 13th century from the Mediterranean (the name comes from a Greek word meaning "sweet root"). In Arthur Young's time from 50–100 acres were grown, and mill-workers used it as a thirst-quencher. Thirty years or so ago John Betjeman wrote:

> *In the liquorice fields at Pontefract*
> *My love and I did meet,*
> *And many a burdened liquorice bush*
> *Was blooming round our feet.*

But the crop was declining, and the last growers, Mr. Booth and Mr. Carter, gave up about 1964. Liquorice for the famous Pomfret Cakes is now all imported, though a little of the plant is preserved by the Corporation's Parks Department.

The other unusual West Riding crop, rhubarb, is still very much a commercial proposition. A few square miles round Leeds and Wakefield contain half the commercial outdoor rhubarb in England (5,000 acres) and three-quarters of the forced rhubarb (733 acres out of 1,012 in 1966; 572 acres out of 741 in 1969). The reasons for this extraordinary concentration of a crop in one area are partly geographical, partly historical. A cold autumn climate is necessary for "vernalization" of the dormant roots preparatory to early forcing. On the other hand good and rapid transport to the main London markets is essential. Leeds is at the point where the southern limit of the one requirement and the northern limit of the other coincide. The heavy moist soils overlying the coal measures appear to suit rhubarb, which also enjoys the high rainfall of the district and even the industrial atmospheric conditions. No other crop could be grown with such advantage over other areas. Low cost fuel from the Yorkshire coalfield was available for heating, and ample supplies of shoddy from the woollen towns (for humus as well as for manurial properties), sewage sludge, and screened town waste and ashes, all of which over the years have built up soils ideally suited to rhubarb. A great deal of skill and expertise goes into the production of forced rhubarb and this has been fostered and handed down from one generation to another—family secrets and special root strains often being jealously guarded.

Rhubarb is not true-breeding from seed, and propagation is from sets—a major segment of root with at least one crown bud. Splitting takes place in winter or spring and 6,000–7,000 sets are planted to an acre. These remain for two or three years in the field before being ploughed out—usually in the third autumn—for forcing. Forcing is done in dark insulated sheds with brick walls and low timber roofs cladded with felt. About half these were built before the last war and are not suited to modern methods and machinery. There are about 160 commercial growers in the West Riding. One or two are very large, with 70 to 100 acres of rhubarb, but half of them grow less than 10 acres. Rhubarb acreage usually forms about a quarter of the whole farm acreage so as to give a long break between successive crops, thus helping to avoid disease. In spite of this, various viruses have weakened many stocks and experiments have been going on at the Stockbridge House Experimental Horticulture Station near Cawood to grow virus-free stocks. Growers have formed a Nuclear Stock Association of which one of the leading growers, Dennis Macaulay of Wrenthorpe, Wakefield, is chairman. Modern systems of oil-fired heating are being adopted but returns scarcely justify the high capital cost of new sheds. The area probably produces 5,000 tons of forced rhubarb a year, worth something like half a million pounds, but both labour costs and capital requirements are very high. West Riding production is greater than that of the United States!

8. *The Northern Pennine Dales*

NORTH of the Aire, Pennine Yorkshire is almost entirely rural. Geologically it may be subdivided into Craven and the Limestone Uplands, including upper Wharfedale, a great deal of Wensleydale, and parts of Swaledale; and the Millstone Grit areas, especially Nidderdale. In fact each dale has its own unity, over-riding geology, and though the heads of the dales are often closely linked—Wharfedale and Wensleydale for example—I propose to deal with each in turn.

Craven is an area of grassy limestone uplands sloping gently up to the north—an excellent grazing country. Much of it lies below 1,000 feet and only to the north do the great fells go up to over 2,000 feet—Fountains Fell, Ingleborough, Pennyghent and Whernside. In former times there was more arable land in Craven than in any of the dales, but for a century now it has been almost entirely grass. From here came the long-horned Craven breed of cattle, of which the most famous representative was "The Craven Heifer" whose picture appeared on notes of the Craven Bank as well as on many inn signs. It was bred at Bolton Abbey in 1807 and sold for £200 when it was four years old. When shown at Smithfield it weighed 18½ cwt. Many of the lower farms in Craven are milk or milk and sheep, while higher up sheep predominate. Even on the hills, especially Malham Moor, there is a long tradition of taking in horses and cattle for summer grazing. William Bray in his *Tour to Yorkshire*, 1777, wrote—:

> From Kilnsey to Malham the ride is truly wild and romantic; nature here sits in solitary grandeur on the hills which are lofty, green to the top, and rise in irregular heaps on all hands, in their primaeval state of pasture, without the least appearance of a plough or habitation for many miles. In the summer they afford good keep for cattle, great numbers of which are taken in to feed from April or May to Michaelmas. The pasturage of a horse for that time is 14s.; a cow 7s.; a sheep 1s. 6d. Many of these pastures, which are of great extent, have been lately divided by stone walls, of about 2 yards high, 1 yard wide at the bottom lessening to a foot at the top. A man can make about seven yards of this in length in a day, and is paid from 20d. to 2s. The stones brought and laid down for hi n cost 7d. more.

A typical farm of the area between Gargrave and Malham Moor

is Winterburn Hall, lying mostly between the 600 and 700 feet contours. Mr. E. Taylor has 300 acres there, including a small amount of enclosed moor, but the land is mostly first class grazing and apart from his sheep (as many as 300 at times), he has a high-yielding herd of 30 Friesians kept in a modern byre (though the Craven word for cow house is in some places *mistal* and in others *shippon*) with pipeline milking to bulk tank.

While sheep in the area are generally Dalesbred or Swaledale, down in Appletreewick in Wharfedale Mr. James Falshaw keeps a "foreign" breed—Welsh Halfbreds—the progeny of a Border Leicester ram and a Welsh Mountain ewe. He says he keeps "what'll pay the rent best," and these suit his slightly lower conditions admirably. He crosses them with a Suffolk ram. On his 150 acre Prospect Farm he also runs 15–20 Friesian cows. The Falshaws were in this district in 1372.

Beyond Grassington, and even into Littondale, there are useful looking fields and farms, a mixture of good dairy herds with sheep and some beef. But where Cray Gill comes into the Wharfe just above Buckden, the valley bottom is 800 feet, and above Hubber-holme there is no bottom land in the agricultural sense of good meadow. At Beckermonds the dale bottom is over 1,000 feet. Langstrothdale Chase was a medieval deer forest, and in the last century there were many leadminers up here and many part-time farms. But there were also many good yeoman farmers who left substantial—and beautiful—stone houses behind them. Farming was never easy up here. Its basis was sheep, with some cattle rearing, and cows for cheese and butter only. In those days most people were content with subsistence farming, wool brought in a fair cash income, and many drew wages from lead mining.

In 1971 there are still fifteen families in the dale above Hubber-holme who depend on farming—and there just isn't the living in farming for them. What can you say to a young man who is trying to live off 150 acres of limestone shelf with 200 Dalesbred ewes and just two or three suckling cows? He cannot possibly make the minimum agricultural wage. Another young couple had 400 sheep on 400 acres of similar land—the rent of which was 25s. an acre— and were trying to boost their income with a dairy herd of 36 cows. But almost all hay had to be bought—and there were no buildings for that number of cows. In winter cows had to be kept in half dozens in scattered barns, and the milking machine carried round. This is slavery. Milk up here is uneconomic to produce and un-economic to collect and take to market. There is one part-time industry which helps to save the situation in these high valleys, and that is taking in visitors. In one of the most beautiful areas of England, there must be several farms where the wife can earn more from visitors than the husband can earn from his sheep. From the strictly farming point of view the only hope at this altitude is to run two or three farms together to make a unit of a thousand acres. Only with a considerable contribution from the tourist trade can smaller farms carry on.

One more controversial solution is afforestation. Against considerable outcry the Northern Pennines Rural Development Board (now defunct) accepted plans for a large acreage of private planting, including a good deal in upper Wharfedale. Amenity and recreational interests as well as sheep farmers have protested, and it is to be hoped that any further large scale afforestation may be prevented or strictly limited to areas where it might possibly be an improvement. There are large areas of wide shallow valleys and lower hillsides of a boggy and peaty character, particularly on the millstone grit, which if drained might produce timber more profitably than mutton and provide more local employment. For any such planting to take place on the fell tops or in the limestone valleys would be a disaster. The protagonists of forestry may talk of landscaping, but a forest which is economic cannot be beautiful. It may be diversified at its edges, but the main part of it must be close-planted and homogeneous. Nothing is more tedious than walking through interminable aisles of pine or larch, in contrast to the varied forests of natural regeneration such as the Caledonian pines of the Cairngorms. Moreover, owing to the system of forestry grants and tax allowances it is very difficult to get a true economic picture. If the same public money that went into forestry were put into farming in the high dales, into providing better buildings and draining and improving hill grazing, farming might show a better return than forestry. As it is, the amenity and recreational interests should have the last word.

Before either farming or amenity interests spoke with such a strong voice, reservoirs were established at the head of Nidderdale. These have toned in with the landscape better than most, and Gouthwaite in particular is a haunt of wild birds, adding to rather than detracting from the tourist attractions. Above Pateley Bridge, Nidderdale is another dale head more likely to profit from tourism than from farming. At the top of the steep hillside above Pateley itself, George Herbert Wardman is the third George Wardman to live in Somerset House. His grandfather built it in 1871 and turned the old house into a cattle-shed. He has 80 acres of intakes—stone-walled fields by the moor edge, won from the heather by previous generations. None are under 800 feet. He also has stray over three or four hundred acres of Pateley Moor—up towards Bishopside and High Ruckles. That is all black ling—heather and no grass. He runs 280 Scotch ewes, putting half to a Teeswater ram to produce Mashams for sale, and half to a Swaledale ram to produce hardy replacements for his flock. He also keeps two or three cows for rearing calves, and sometimes buys in Friesian heifer calves for rearing. He sells a few down-calving Friesian heifers every year. But the summer is short and the grass is scarce here. All the intakes together will scarcely produce as much as a good ten acre field on the plain. "Tha hes ti be born ti this sort o' spot an this sort o' life," he laughs—and he himself, with his three-year old grandson helping him to shepherd his lambs and fodder his calves, looks firmly and happily established on this difficult spot after a century of family occupation. But he is near retiring age — if old farmers ever do retire! It is almost certain that his grandson will have nostalgic memories of this kind of life which is now so rapidly disappearing, but will never farm himself in this way. He, or someone, will continue to live in this beautiful place, looking across the deep valley to Greenhow Hill and the sunset. A few calves may be kept as a hobby. Unless the intakes and the moor go to join a bigger unit, the more distant fields may well go back to moor.

Below Pateley, however, there is good grass farming in Nidderdale. Hartwith, on the northern side of the dale, sloping down to the south and the sun, was where my own love of farming began in 1922. I was seven; we lived for a year in the old grey-stone vicarage and I went to the village school. In fact there was no village at Hartwith— just the church and the school and the scattered farms. But it was a close-knit community in which most families had known each other for several generations. All met at church on a Sunday morning, and stood in groups in the churchyard afterwards exchanging the news of the week. You could always see across the dale exactly what was going on in all the fields on the south side, in Dacre and Darley, but you couldn't see much of the fields on your own side! It seemed a golden time then in that farming world, to which all who belonged to it look back with the same nostalgia. There were choir practices, sales of work, a church trip by charabanc to More-cambe, a 5th November parkin party round the bonfire. No-one is left now in Hartwith of that community of only half a century ago. Fryers, Lumleys, Hastings, Housmans, Fawcetts, Robinsons, Lambs have all gone.

As a boy I soon found my way over the stone-stepped stile to John Fryer's Prospect Farm. Hartwith lies between the 500 feet and the 700 feet contours in the main, going up to the north to 950 feet at Brimham Rocks. But below 700 feet a mixed farming system was practised then. I tried my juvenile hand at milking, butter-churning, calf-feeding, bullock-chasing (and bull-avoiding!), muck-leading, turnip - cutting, and above all hay-making. Those high, breezy meadows where in memory the sun always shone (it must have been a fine summer that year) made sweet hay—all turned by hand with a rake. Then the hay was put into big "pikes" (not small hay-cocks here) and swept with a horse-sweep to the stack or loaded on coup carts complete with hay-shelvings to be put in the barn. At 10.0 a.m. and 3.0 p.m. "drinkings" were brought to the field—a can of tea, cheese and bread and apple pie—and a moment of leisure enjoyed in the shade of the stone wall among the fine grass and the harebells. There were a dozen or so acres of corn to harvest too—and I helped to tie bands behind an old manual reaper where one man drove the horses and another with a hand rake put the bundles of corn off the right size for a sheaf. Of course there were self-binders in plenty on the plains then, and even combines had been heard of in distant parts—America perhaps. But not many people of my age have worked behind a "manual." Just thirty years later I bought my first combine.

Hartwith today is practically all grass, and good grass. The 83 acres of Prospect Farm are run together with the similar sized farm of Edge Nook (this was an inn till the 1890s, and when John Fryer's eldest son Leonard put electricity in he put the engine down in the beer cellar). Mr. R. M. Burrow also rents another 60 acres and has an 80 cow milking herd loose-housed in some very large sheds. Bedding is all to buy in however, and in 1970 straw and shavings cost £500. There is a scheme to put in yokes or cubicles to save most of this expenditure. There is a 300 gallon refrigerated bulk milk tank. Nothing could be more of a contrast to the Hartwith of the 1920s. The annual gross income of this farm must be three or four times that of the whole parish in that distant era.

Even more than Nidderdale, Wensleydale has long been famous for its pastures, its milk and its cheese. Ewes' milk cheese was made by the monks of Jervaulx Abbey, whose recipes remained in use after the Dissolution. It was probably not till after 1800 that the shorthorn cow displaced the ewe as the main source of milk for cheese. In 1794 Tuke ranked Wensleydale as "the foremost of the dales, both for extent and fertility; the bottom consisting of rich grazing grounds. In Wensleydale scarce any land is in tillage." Yet in 1800 Leyburn was considered one of the largest corn markets in the north—perhaps because of corn being sold there from the lower to the upper valley. East of Leyburn the lower parts of Wensleydale and all the rich land round Bedale and Masham may be considered for agricultural purposes almost part of the Vale of York. Although towards Masham there are still a lot of sheep, there are also some large arable farms and some very large dairy herds, as well as intensive pig and poultry units. My own farm is linked with some of these near Bedale in a Farmers' Purchasing Group—bulking our orders for fertiliser, feeding stuffs, and other farm requisites in order to get the lowest prices. There is also some trading among members.

In Wensleydale proper sheep are still very important. The hornless Wensleydale breed was developed by Thomas Willis about 1838 from a Leicester ram "Blue Cap" on Teeswater ewes. The Willis family (who still farm at Carperby) have done a great deal for both sheep and cattle breeding in Wensleydale. They still have a small flock of traditional Wensleydale sheep, though most people now seem to consider Wensleydale and Teeswater as much the same, and either ram on a Swaledale ewe will produce the very popular "Masham," much in demand for cross-breeding even outside the county. Some years ago the Smithfield fat lamb championship was won by a pen of Hampshire–Masham crosses from a farm not far from mine. The Willis family played a big part in shorthorn breeding. In early days these were the beefier type and one beast from Leyburn won the Supreme Championship at Smithfield and was then roasted whole on Boxing Day and distributed to the poor of Chelsea.

The main product of Wensleydale, however, remains milk—and cheese. Earlier in this century a "milk train" went to London daily with 6,000 gallons of milk. Farmhouse cheese-making was important then, though "factory" cheeses began to be made in 1897. Edward Chapman, who had collected cheeses from the farmhouses to take to Yarm Fair every October, started a factory at Hawes in an old woollen mill and was soon followed by Alfred Rowntree at Masham, Coverham and Thoralby. For many years now only one or two individual makers have been left, in the highest parts of the dale. Even factory cheese-making was threatened in 1935—the new Milk Marketing Board were going to close down the Hawes factory. That well-known dalesman, Kit Calvert, persuaded, cajoled or blackmailed sufficient farmers to get together to raise a starting capital of £1,085 in £5 shares. Kit wouldn't promise to take milk unless a farmer bought two of them, and he put up £100 himself. A local catastrophe was averted, and Wensleydale cheese saved for a nation that has proved grateful for it. In July 1966 the Wensleydale Creamery sold out to the Milk Marketing Board for £485,000.

In 1935 the Creamery had to guarantee 500 gallons of milk a day. Recently they have been handling 4,500 gallons at Hawes, 7,000 at Kirkby Malzeard, and 1,000 gallons at Masham. A good deal of this is sent on in bulk at seasons of liquid milk shortage—the dale thus acting as a sort of milk reservoir—but overall cheese production has not suffered and during 1966 Hawes for the first time turned out over a quarter of a million of the 1 lb.miniature cheeses. Production has also started of the traditional Blue Wensleydale, one of the finest of all cheeses, the recipe for which was thought to have been lost. It is produced at Masham—which also boasts one of the few small local breweries left, Theakstons. There is an Express Dairy at Leyburn which handles liquid milk entirely, exporting in bulk to Bradford, Hull and Liverpool. Amounts vary from 12,000 to 18,000 gallons a day according to season. It is estimated that altogether the heart of Wensleydale, from Masham to Hawes, produces 20,000 gallons of milk a day, more at the peak. And at the peak Kirkby Malzeard and Hawes make a good deal of butter too.

The man who saved Wensleydale cheese, Kit Calvert, was born in Hawes in 1903 and started work herding cattle in the mart for 5s.

a week. He was soon dealing in cattle, and renting land to keep his own cows on. He got 8d. a gallon from milk in winter and 5d. a gallon in summer from the Chapman creamery. After he had rescued this from closure in 1935 he had to fight the Ministry of Food during the war, and thanks to his blunt speech Wensleydale alone of territorial cheeses was allowed to continue in a small way. But Kit is not only a farmer, cattle-dealer, business man and designer of modern creameries. He also owns a small shop full of old Yorkshire books, and is steeped in the history and dialect of Wensleydale. The dialect is his natural native speech and he has translated into it several chapters of the New Testament. He has never left the dale except for brief business trips. Another Wensleydale man, John Blades from Lunds, went to London in 1779 with half a crown in his pocket and in 1813 became Sheriff of the City (like that other dalesman, William Craven of Burnsall, who was sent to London by common carrier in 1548 to be apprenticed to a mercer and who in 1600 became Sheriff and in 1610 Lord Mayor).

Above Lunds the Yorkshire boundary goes up the deep narrow chasm of Hell Gill onto Ure Head and High Seat. Across Cotterdale is Great Shunner Fell—all wild difficult country. Here are the great Commons of Abbotside and Angram. Abbotside—so called because of its connection with the monks of Jervaulx—was a stinted common from 1837, looked after by four shepherds, for whose wages all common right holders had to pay an assessment.

The problems here, like the moorland grazing, are shared with Swaledale, the most rugged of all the dales, and the most essentially dependent on sheep. In this steep-sided and narrow dale the proportion of good bottom land to fell grazing is low, and the dale suffers more than most from the two special problems of the high Pennine farms—the shortage of winter keep and the high cost of transport. A wagon load from the plains of York or Cleveland to Keld now costs £10–£12, adding £2 a ton (apart from extra handling) to the cost of straw. Every autumn we take in on our Cleveland farm a hundred or so lambs from the top end of Swaledale, lambs that have been reared on the bleak fells of Birkdale Common and Nine Standards Rigg—running right into Westmorland and up to 2,000 feet. They do well on our surplus grass, eating down tufts and patches left by cattle, and going off fat before severe weather sets in after Christmas. Our Swaledale friends who bring the lambs down on agistment buy hay, straw and corn from us to take back up there.

The Whitehead family farms some of the remotest farms in Swaledale including till recently Raven Seat, up Whitsundale, which is two miles away from anywhere at all! George W. and Laurie Whitehead are at Pry House, Keld, 1,300 feet above sea-level. Even there until recently they had a small milking herd of very good Shorthorns but are now out of milk, finding that suckler cows, with the Hill Cow subsidy, give a comparable or better return for less labour. There are 42 cows here and at Kisdon Farm just below Keld which is shared with another brother, Ernest. There are 300 ewes and followers on Birkdale Common and 100 on Kisdon. But Laurie Whitehead also buys in a lot of lambs and may have as many as 1,500 away on agistment at various plains' farms. Other brothers farm at East Stonesdale, Swale Farm at Muker, and Summer Lodge at Low Row. They would probably tell you that they were only two crosses removed from the Swaledale sheep that are their main interest and main livelihood.

Tuke in his *General View of the North Riding* in 1794 described the sheep bred on the "Western Moorlands" as "horned, having grey faces and legs, and many of them a black spot on the back of the neck, wool rather coarse and open." It is generally held that the breed has changed considerably even in the last fifty years, towards a longer and finer fleece, but one that "fills the hand well." The type now is a black face (still sometimes grey or mottled) with a white nose. I still notice the black spot behind the head on some of the lambs I handle in the autumn. The wool is not so long as that of the Rough Fells from further west, towards Ravenstonedale and Kendal, which makes more per pound than the Swaledale wool. But the Swaledale is reckoned a very hardy breed which is infiltrating into the Cleveland moors and even into Scotland. A great deal to improve the breed has been done by the Swaledale Sheep Breeding Society, and there is always the hope of adding a little to the income on the higher farms by breeding some good pedigree rams. A good ram might bring £200, and even £500 has been known. In contrast Swaledale lambs sold as stores (for fattening) do well to make £3.60. The average killing out weight of a fat lamb is 26 lbs., at 20p. per lb. —say £5.20–5.50 for a lamb. Four-sheer ewes—that is ewes after their third lambing—whose teeth are beginning to get too worn or broken for them to do well on the tough moor grazing, are drafted out to the lowlands at somewhere round £5 or £6 per head for crossing with Teeswater or Blue-faced Leicester tups to produce Masham lambs. A Swaledale fleece is worth 60–65p. On the fell, Swaledale ewes average less than one lamb per year — there are always some losses of ewes and of lambs, and a fell ewe cannot support twins, which have to be fostered onto a lambless ewe or reared as "pet lambs." With the hill sheep subsidy—now £1.60 per ewe—the annual profit is, as in Cleveland, about £3 per ewe— probably slightly more. It would be a struggle, even in a favourable year, to make £4.

Conditions are hard on the higher farms, where it is usually reckoned that winter lasts nine months. Perhaps because of the hardiness of the Swaledale breed, there has however been no real disasters since the bad weather of 1947 which halved the flocks. The biggest of the fell flocks is probably about 600 ewes. Without the hill subsidy life would certainly be grim; one must look on the subsidy as some compensation for the extra cost of living and farming in these high dales, and probably not enough compensation at that. With all his enterprises Laurie Whitehead may make as much as a London docker or a Midlands car-worker—but he works a 12–16 hour day seven days a week to do it.

Tuke in 1794 was most impressed by the method of hay-making practised in Swaledale for "the speedy and equal exsiccation of the grass." The swathes were quickly and frequently strewn about, "but always with the hands, not forks, the miserable invention of indolence. Hay is the grand object of the farmer and he bestows on it the most sedulous attention." Swaledale is still primarily concerned with making good hay, and there are still farmers who like to get among it with their hands. But below Grinton a fair amount of barley is now grown, where once there was very little above Richmond. Richmond itself is almost at the beginning of the plain, and even has the traditional offering of a bottle of wine to the first farmer that brings along a fair millable sample of wheat from the year's harvest.

The higher dale is as Alpine as any part of England, and the land itself defies modernisation and mechanisation. The Landrover, the tractor, and electricity can help, but they cannot here transform an economy. Success still depends very largely on sheep and on shepherding. There are however some modern problems. The traditional barns or laithes scattered round the fields are less and less used for farming purposes. There are seven on the Whitehead's Kisdon Farm alone, lovely stone buildings which are tending to deteriorate. Yet the National Parks Planning Authority will not allow them to be converted for habitation. Cottages and landless farmhouses which fall vacant are usually bought by professional men from the towns at higher prices than local people can afford, and used as holiday homes. Young couples belonging to the dale can sometimes not find a home there. Some of the smaller and more isolated farms where living is really a struggle are being given up, yet there is a limit to what one of the bigger enterprises like the Whiteheads can take on, both for capital and for labour. Practically no casual labour is available in the dale nor is there any accommodation for any. No-one wants to see the essential character of the dales changed, or any new development which is not in complete harmony with the old. But there is a real danger of the planning policies pursued hitherto leading to dereliction and sterilisation, with half the houses occupied only at weekends and walls and barns falling to ruin. A converted barn would be better to look at than a ruined one. To allow strictly regulated conversion of some of the surplus stone buildings in the dale would mean providing more accommodation, bringing more people and money into the dale, and converting useless buildings into much needed capital for improvement of dales farming.

There is another cause for concern among dales farmers. Some of the bigger landowners are trying to reduce the number of sheep on the fells and are buying up stints on the big commons or objecting to the numbers of sheep registered under common rights. Their aim is to remove some of the competition for the fell from the grouse, since shooting rights can sometimes now be worth more than sheep grazing rights. In the Swaledale area there has been letting to Italian and American syndicates. There is danger of another kind of sterilisation here, such as has closed off large areas of the Scottish Highlands for deer-stalking. All these trends together form a serious threat to the economic and social life of the dales, already in a state of flux.

Traditionally, Swaledale also made a cheese which is said to have been better than that made in Wensleydale. One or two cheese makers lingered into the 1940s—it may be that even now an occasional cheese is made for home-consumption, as in Dentdale and Teesdale. I have in recent years tasted a Teesdale or Cotherstone cheese—a creamier texture than the average Wensleydale. Historically, farm incomes in this area were supplemented by the universal cottage-industry of hand-knitting. The "terrible knitters of Dent" were famous and in Swaledale a man might sit down for a rest and a chat for "six needles"—the time taken to knit six rows.

The Yorkshire part of Teesdale has similar problems to Swaledale. Both dales had a great influx of lead miners in the 18th and 19th centuries. With other mineral workings and quarryings, they left some farming problems and some dereliction behind. The Dales are still historically recognisable—but how long they will remain so is a matter of grave doubt.

9.   *The Future of*
                       Yorkshire Farming

IN the years ahead farming on the Yorkshire lowlands will share the same problems and follow much the same course as farming elsewhere. The move to very big farms may be slower here because of the strength and tenacity of the comparatively small—though not the very small—family farm. By the time the present farmers of 60 and over have retired, any farms of 60 acres and less that survive will be part-time holdings (excluding very intensive "factory farms" and market gardens). The 100 acre farm on good land, probably with an intensive dairy herd of 80 cows, run as a family unit, will survive for the foreseeable future. As a general rule the mainly arable farms will have to be at least 300 acres to survive—perhaps in ten years time they will need to be 500 acres to spread the higher cost of machinery. The large farming families, co-operating among themselves on several farms, will continue to strengthen themselves and build up their holdings. The greatest danger to them is the impact of death duties and Capital Gains Tax, although the latter has been eased a little recently.

The family farm and the farming family, stable over the generations, is the best reservoir of health and character that a nation can have. Farming at its best is a very long-term business. Over the past century we have been too dependent on cheap imported food. The next half century may see great changes in that. A nation that neglects its land or taxes its family farms out of existence may suffer greatly for it. No taxation based on inflated land values should be imposed unless land is actually sold. With land prices four times the level of even 25 years ago, a quite modest hard-working family enterprise of 200 or 300 acres can be crippled, probably ended, by two deaths within fifteen years, a by no means uncommon occurrence. This cannot be in the national interest.

In the dales and on the moorlands of Yorkshire the problems are much more serious, the future very much more doubtful. I had not realised, until I did this survey, how far the changes had already gone. Some dales, such as Bilsdale, have already almost completely altered their population and their character. The Pennine dales are altering fast. Here and there a young dales couple are setting off bravely to farm in the family tradition against economic odds and ignoring all the attraction of "civilised amenities," but they are few and far between. In the upper dales their only hope would appear to be in providing accommodation and catering for visitors—and they get little help or encouragement in that from any source. There is urgent need for a survey of this kind of catering, much in demand by visitors to the National Parks—such an examination as might be given to any other farming enterprise, of the best ways of doing it and the right prices to charge. In comparison with rapidly increasing hotel prices, farm-house charges for meals and accommodation probably remain too low.

There is also great necessity for a study of ways of increasing accommodation. Large caravan sites should certainly be kept out of the dales. But there are many farms where a small group of ten or twelve caravans could be sited in a secluded position or screened with trees, and such a group could add £300 a year to a farm's income from site-rents alone, making all the difference to its survival. It is a great pity that caravan design has been too much affected by the completely erroneous assumption that caravans are going to be used for travelling. This will be increasingly difficult, and more and more of them will remain on fixed sites. For this the design could be much more pleasing, such as one flat-topped kind of mobile home made in Yorkshire. Or for dales farms some stone and wood chalets might be designed that could fit admirably into the landscape. What is vitally necessary is for County Councils and planning authorities to take a much wider view of the social and economic factors involved, and not look only at the landscape. If all development of this kind is to be prevented, then some sort of special National Parks subsidy should be paid, of at least £400 per year per farm involved, in compensation. But it should be clear from this brief review that to keep these dales as museum pieces is the one certain way of killing them stone dead. Unless dales farms are allowed to cash in on tourism in some way, or realise capital from buildings now agriculturally useless, other strong economic forces already at work will lead to the dales becoming sterilised areas of holiday cottages, afforestation and game preserves.

Indeed the dales and the dalesmen as they were are already doomed. Perhaps the process was inevitable. We have had the Wild Life societies preaching conservation—"People need pandas." It seems to me that we needed the old type Yorkshire dalesman a great deal more. But he is disappearing fast. Few will be left in another ten years. All that reservoir of wit, character and folk-lore will have gone, the blunt independent small farmers, full of dry humour, heroic in mould and poetic in speech, hard-working but deeply contented, part of the scenery and of history. Even farming is a rat-race now, and only the most economically efficient will survive. Some will be dales people, and they will remember their fathers, but they will not be like them. As one farmer and miller Basil Cockerill wrote:

It's middlin strange thoo nivver knooas
What thoo likes best
                                        Until it gooas.

There was neea prooder married chap
Nor me, wi t'gallowa an trap,
Drivin on a winter's neet,
Trap lamps lit bi cannleleet.
Creeakin, jinglin on t'high rooad,
(Under t'rug were nooan so caud).
T'awd mare, Dolly, allus knew
When we'd reached t'last mile or two.
Unexcitin, doonreet plain—
Back wheer we started,
                                        Doon oor lane.

At winter's end. Upstairs i t'dark
I hear t'owls call an t'dog fox bark.
Then t'awd man dreeams he's young ageean,
An Dolly clip clops
                                        Doon oor lane.

It's middlin strange, thoo nivver knooas
What thoo likes best
                                        Until it gooas.

Bibliography

Young, Arthur, *A Six Months' Tour through the North of England* (1770).

Marshall, William, *Rural Economy of Yorkshire* (1778 and 1796).

Tuke, J., *Agriculture of the North Riding of Yorkshire* (1794).

Leatham, I., *General View of the Agriculture of the East Riding of Yorkshire* (1794).

Brown, R. (and others), *General View of the Agriculture of the West Riding of Yorkshire* (1794).

Edwards, W., *Early History of the North Riding* (1924).

Elgee, Frank, *Early Man in North East Yorkshire* (1930).

Pontefract, Ella and Hartley, Marie, *Swaledale* (1934), *Wensleydale* (19—) and *Wharfedale* (1938).

Harwood Long, W., *A Survey of the Agriculture of Yorkshire* (County Agricultural Surveys) (1969).

Raistrick, A., *The Pennine Dales* (1968).

Yorkshire Archaeological Society (Helmsley Group), *A History of Helmsley, Rievaulx and District* (1963).